Selling Commercial and Industrial Construction Projects

Selling Commercial and Industrial Construction Projects

W.D. Booth

VNR VAN NOSTRAND REINHOLD COMPANY

NEW YORK   CINCINNATI   ATLANTA   DALLAS   SAN FRANCISCO
LONDON   TORONTO   MELBOURNE

Van Nostrand Reinhold Company Regional Offices:
New York Cincinnati Atlanta Dallas San Francisco

Van Nostrand Reinhold Company International Offices:
London Toronto Melbourne

Library of Congress Catalog Card Number: 78-16643
ISBN: 0-442-20899-5

Manufactured in the United States of America

Published by Van Nostrand Reinhold Company
135 West 50th Street, New York, N.Y. 10020

Published simultaneously in Canada by Van Nostrand Reinhold Ltd.

15 14 13 12 11 10 9 8 7 6 5 4 3 2 1

Library of Congress Cataloging in Publication Data

Booth, W   D
    Selling commercial and industrial construction
projects.

    1. Construction industry–United States.
2. Building–Contracts and specifications.
3. Contracts, Letting of–United States. I. Title.
HD9715.U52B64   658.89'69   78-16643
ISBN 0-442-20899-5

Introduction

With a jolt you pull your car to the side of the road, and read the job sign that can't be over one day old: New Home of ABC Inc.; General Contractor, XYZ Construction Co.

You think out loud, "I knew the property had been sold," and your eyes roam over the real estate sign with the angled "sold" sticker proclaiming to all that the agent has been successful. "Thought the damn job would have come out on the bid list by now," you mutter as you move back onto the road and head for your office.

Later in the day you get the agent on the phone: "Jim, how did XYZ get the ABC job? I saw nothing on the builders' exchange about it."

"John, XYZ has been negotiating with ABC for the past six weeks. They did one fine job of selling on old McDuff. Understand they just called on him out of the blue asking for a chance. They put the whole package together, plans and all."

"Didn't that tightwad McDuff get another price, Jim?"

"Sure did; two, as a matter of fact. He asked me to get two other contractors capable of putting the complete project together. He liked XYZ better. I don't know the prices, but I really feel those boys impressed McDuff by going to him. Were you planning to bid the job?"

"Thinking about it," you answer, trying to sound very casual.

"Well, I wish I'd known you were interested. I'd like to have recommended you to old McDuff, but I thought you mainly bid work off the exchange. I'll call you next time I have a buyer who wants to negotiate."

"Please do. Say, Jim, how about eighteen holes this Saturday at the club . . . ?"

I'm sure this has happened to every general contractor at some time; but maybe it's occurring too often, and such a disturbing piece of news has prompted you to investigate the design-build-negotiate avenue of getting business for your firm.

The problem is, you are puzzled about just how to get started; after all, you aren't selling cars or TV sets but buildings, and buildings aren't sold like cars and TV sets! Or are they?

The idea of selling construction is the innovation that progressive general contractors are just starting to realize ranks with the change concrete brought about in the industry when it was widely accepted years ago.

The purpose of this book is to give guidelines and answers to questions on how to break into this field or to improve efficiency if you are already negotiating business. Some suggestions may go against your set methods of operation, and you may be reluctant to follow them; but at least consider giving them a try. Chapters 1 and 2 cover selling construction in a broad and general way. The attitudes of both the contractor and the customer are discussed. In short, the first two chapters will give you a feel for what you are letting yourself in for, while the rest of the book covers the bottom line nitty-gritty of obtaining results and making money.

I have no idea how or when selling construction started. I don't think anyone knows. For my money, though, it was the metal prefab buildings segment of the industry that adopted the idea of going to the potential customer and asking for business. The idea of selling a building has been very successful for the metal prefab people, and the same principle can be applied to conventional construction as well.

The guidelines in this book can be utilized for any size project, be it $500 or $5,000,000. But, from my personal experience, the area that offers the highest percentages of contract signings is between $20,000 and $250,000. Let me qualify that statement. As general contractors doing commercial and industrial work, you know any job under $20,000 is not much of a job. The exception is a small add-on or remodeling project. The small job is usually wanted by a small operator who does not have much capital; hence it turns into a nickel-and-dime contest. As for the large jobs, large private concerns usually follow the bid route, with full plans prepared by an architect.

The area between these two extremes is ideal for the design-build-negotiate concept. This segment contains the independent businessman whom you can approach in a one-on-one situation and sell your services, sell saving money, sell saving time, while maybe helping the customer get the jump on his competition. You sell the job, not bid it. I'm not saying that you'll have no competition, as other construction companies may be after the same contract. What you have to do is convince the prospect that out of all the people he is talking with, you are the one from whom he will get the best deal; and quite often this does not mean the lowest price.

One personal by-product of selling construction should also be mentioned—the excitement and just plain fun of putting a deal together and making it work. Let me put it another way. Bidding is like airplane bombing: you fly over the target and drop your bid, and some time later, back at the base, you hear if you hit the target. On the other hand, design-build-negotiate takes place on the front lines where business is conducted eyeball to eyeball, which is an exciting way to make money.

Every book on selling I've ever read does nothing but state broad principles about salesmanship. The small but pertinent questions concerning the details never are answered. Matter of fact, they aren't even asked. I have slanted this book the other way, and small details will be the norm, not the exception.

I will not try to give figures for you to follow on what should be made or what to charge an owner. We are mainly interested in selling the job in this book. Working knowledge of the economics of a construction project is a must, and I assume you know your way around in that area.

I'm sure as can be that someone can challenge every statement I make in the following pages, and back it up. I don't wish to take issue with anyone; all I'm saying is that the information in this book has worked extremely well for me.

There are no guarantees! When all is said and done, it will still be up to you to make things happen.

W.D.B.

Acknowledgments

I would like to express my sincere thanks to the following:

Warren H. McNamara Jr., Attorney at Law, of Hampton, Virginia for his help and timely advice with certain sections of this book.

Walter E. Carter of Virginia Beach, Virginia and R. B. and C. J. Lindemann of Norfolk, Virginia for allowing me to include copies of actual contractual documents.

And most of all to the one truly indispensable person, my wife Julie, who did all the typing.

Contents

Selling Commercial and Industrial Construction Projects

Chapter 1

The Art of Selling to the Construction Prospect

Selling is an art, not a pure science. It is dealing with people; thus there are no ironclad rules such as one has in the sciences. The rules change with the individual, so the person practicing the art of selling has to be able to adapt to the situation.

Selling is simply one person setting out to convince someone else to buy his wares. Everyone who is in the process of earning a living sells. It can be the man behind the counter of the local hardware store selling tangible goods, or it can be the insurance agent selling intangibles. The secretary taking shorthand is selling skills. The doctor thumping your chest is selling knowledge. The lawyer sitting quietly listening to the details of a potential lawsuit is selling services. When the general contractor puts in his bid, he is trying to sell his ability to perform the work, at the lowest price. This is selling as thought of in its broadest sense. A great many people do not think of themselves as salesmen pushing their goods and services because this idea may not be an accepted part of their way of earning a living. The term that I like to use for the selling of those people who sell but who don't recognize the fact that they do is "passive selling." Members of the professions, office workers, military personnel, factory employees—these are all passive salesmen, making up by far the bulk of people in the job market.

The other side of the coin is represented by the person who accepts the fact that selling is one of the most important business functions that he can perform—the aggressive salesman. And I don't mean the "foot in the door" type of aggressiveness, but rather the "go to the customer and ask for the business" type. A great many people practice the art of aggressive selling. You see them every day: the building supply salesman who calls on customers regularly, car dealers, shopping center merchants, and general contractors—all are out in the marketplace practicing aggressive selling. They are trying to get the customer to spend his money with them. The result of this effort is the greatest success story the world has ever known: free enterprise.

The express purpose of this book is to help the general contractor bring his aggressive salesmanship to its highest peak by going out in the marketplace and successfully selling his skills, knowledge, and abilities to an individual in need of those services.

Now this is an entirely different type of selling from what the general contractor is used to. Most of his background and experience has been in bidding for business, but now we are talking about going and asking for the chance to build a new building. The person who does this, be it the owner or one of his employees, will have to become a true, honest-to-goodness salesman in every sense of the word.

ON AND OFF BUTTONS

The first point a salesman must understand is that people have *on* and *off* buttons, and knowing how to push the on button is the first step to ending up with the signed contract. At the start you are not worrying as much about the on button as you are about taking pains not to push the off button. For example: I'm sure you have at some time walked into a clothing store, let's say, looking for a new sport coat. You quickly find the sport coat section, and about that time the clerk comes up and asks to help. You tell the clerk that you'd like to browse for a bit; the clerk proceeds to ignore your request and hovers at your elbow telling you how this coat or that coat is made for you. After a few minutes of that pushy selling approach, you leave—not because there was a poor selection of sport coats, but because the clerk pushed your off button.

Think back over some of the times a salesman has turned you off because of the manner in which his approach was handled. Now think about the good experiences, when you have said to yourself, "I like that person, and he knows his business." Chances are, the differences are in the sales personalities of the individuals you are dealing with. The way a person comes across to you at the start creates an impression which is almost always permanent.

When you call on a prospect, no matter whether it's a cold call or one that was recommended, you want to create a good impression. You want to come across as a nice guy, thoroughly grounded in your business, who thought enough of Mr. X's business to come in and ask for it.

I would like to mention right here that whenever possible you should call on the phone for an appointment. This usually happens when you find that Mr. X is in the market for a new building, and you want to contact him. This phone call does two things for you; it shows Mr. X you run a businesslike firm, which is helpful in having Mr. X start to form the correct impression of you; and the call also lets you find out if Mr. X does indeed plan to build, which helps you make better use of your time.

Now I recommend that when calling a perfect stranger on the phone for the first time you state your name, your company name, and then ask if he has a few moments to talk to you. One of the rudest assumptions a person can make is that the other party has time to talk just because he's on the other end of the line. So ask if it's convenient, and if not, say you'll be glad to call back. And ninety-nine times out of a hundred, Mr. X will tell you to go right ahead. So you've cracked the door, and Mr. X hopefully will start to form that favorable impression of you, since, by asking for his time, you have told him, "I know you're a busy man and we're having this conversation because you want to." You have put Mr. X in charge, and everybody likes to be there. After the preliminaries, get to the reason for the call and then say goodbye.

Many times it's not practical to phone ahead, for example, when calling on a likely looking prospect whom you know nothing about. But it makes no difference whether you have an appointment or are making a cold call; the idea is to not push the off button.

When you have an appointment, be on time! If for some reason you cannot make it, call as soon as you know and do give a reason. Mr. X is in business and has problems, so he'll understand when you tell him a truck backed into a freshly laid block wall and you have to be there.

Now remember the only goal you have set so far with Mr. X is not to push his off button. You are in essence simply selling yourself, and you are the first thing Mr. X has to buy, long before he signs the contract.

You arrive at Mr. X's place of business, report to his secretary that you have an appointment, and finally meet Mr. X face to face. But hold it a minute—suppose Mr. X runs an auto parts store, and there are five customers waiting. Then you stand to one side out of the way until Mr. X has everything under control, and wait until he can see you. Then he may leave you to wait on a customer. Mr. X will always put you second to the customer if he is operating a walk-in type of retail business—a fact you'll have to live with. The point is, make yourself fit into whatever scheme Mr. X is operating under. Sometimes it's not easy, but it is necessary.

Regardless of what you have done to meet Mr. X, you are finally there; and, as is customary, the two of you shake hands. Now ask yourself: What did Mr. X think of my handshake? This is important because I firmly believe more off buttons have been pushed by "limp fish" handshakes than all the other reasons combined. If you don't know what kind of handshake you give, then I seriously suggest you find out. Nothing makes a better impression than a brief, firm handshake.

Now that the meeting is under way, you, the salesman, will have to do the initial talking to get the interview off the ground. After all, you've come to Mr. X, and he wants to know the details of why you are there. As the meeting progresses, the good salesman starts to control the conversation. Note that I said control, not dominate. People seem to have the preconceived notion that the hotshot salesman is a smooth, fast talker who never gives the buyer a chance to say a word. Not so!

The really successful salesmen are the best listeners. The first item of importance in the interview is to learn. You need to find out what the prospect's plans are—does he have a site, has he obtained financing, can you help in any way? Are there any special problems that must be looked into? Learn all you can; you can never have too much information about a prospect. All this evolves by the salesman's asking the correct questions—and then sitting back and listening. The impression made is one of trying to be helpful, not one of telling the prospect that you have just the thing for him. Rather, you're asking what the requirements are and how you can help by using those requirements.

By the way, never, never, tell Mr. X how to run his business; that's a sure way to push his off button.

LOW PROFILE AND SOFT SELL

At this point what the salesman is doing is maintaining a low-profile, soft-sell attitude that does not put off the prospect. The salesman is selling himself and making Mr. X start to think about doing business with him.

Low profile, soft sell is the exact opposite of the obnoxious, loud-mouth, contract-waving, slick talker. Low profile and soft sell means doing your selling in a courteous, respectful, businesslike manner. You never presume to tell the prospect what he needs until you have all the facts and he asks you for your opinion. You are helpful, and you don't get bent out of shape when you are inconvenienced by the prospect because of the pressures and responsibilities of running his business. You understand the prospect's particular problems and are able to fit yourself into his picture. You sell to the prospect in the same way you would want to be sold to if the roles were reversed.

Many useful things can be found out during the first meetings with the prospect, but most of them are technical and will be covered in detail in later chapters. Right now we are interested in the general, overall selling picture and how it works for the construction salesman.

The construction salesman has one important thing working for him that puts him right at the top of his profession: he is selling goods that are not purchased unless they are needed. He never pushes something off on the customer. When he sells a building, it's needed. This gives the construction salesman a mental attitude about his work that allows him to think of himself as truly a professional in his business community. The reason is simple. Mr. X might buy some ten-dollar gimmick for his car when he's not sure he needs it; but he'll never buy that 5,000-square-foot storage building unless he really needs it, no matter how good the salesman is. The construction salesman provides the same services as any other professional; he fills the needs of the buyer when the buyer is in the market for his special services.

Now so far our selling attitude has been centered around the idea of low profile, soft sell, but the trick is to achieve this and remain aggressive. The salesman must know just how far to carry a subject, when not to pursue a certain avenue; and this knowledge can only come from experience. Every good salesman has to work out his own system.

The main way to remain aggressive while practicing low profile, soft sell is to show interest. There are many ways to show interest during the selling period of contract negotiations, and those ways will be discussed in detail in later chapters.

TIME INTERVAL

There is an amazing fact concerning the businessman that may or may not be known to you: Mr. X will finally decide to build that new facility only after he is forced to face the hard fact that he must have a new building. He will hang on to his old place for dear life until the very last minute.

For a businessman the prime mover is always money; in every decision it is the predominant factor. Why should Mr. X build the new warehouse and double his monthly mortgage? He will build when forced out by redevelopment, or if a new highway is going through his loading dock; business expansion is always a long agonizing undertaking. The reason I mention this now is to point out that during the selling process it may take months or even years for Mr. X to get to the point where he is ready to sign up and get the ball moving. Construction salesmen must accept this fact. Very few sales are made quickly; most take between two and six months. And if you are doing your job effectively and efficiently, this is the ideal time sequence. The idea is to make contact with the prospect in the initial planning stage and carry through to contract signing, then on into the building stage. Since most businessmen move cautiously, it will take time to do the selling job right.

A quick project, that is, one where the prospect has done a great deal of preliminary work and wants to get started right away, is a project that is already out on the street. The end result is one that boils down to the low price, and that is what you are trying to get away from by selling construction projects.

Obviously, then, for the first months you or your salesman is operating, there will probably be no contracts coming in. It will take time to get prospects in the pipeline so that there are several being worked at various stages; then before long the repeating process will take over, where jobs are dropping out as new prospects are added. Just bear in mind that if you end up with some fast business, it will be more luck than anything. So don't be discouraged when you start selling and get the idea that you are doing nothing but spinning your wheels.

EMPATHY

A good construction salesman should also be able to practice *empathy.* Empathy is the ability to put yourself in someone else's place and see things from his or her point of view. From the salesman's side it may look like Mr. X has no choice but to build, so you get in gear and go after the job. Then Mr. X tells you why he's not going to build; and unless you can change places with him, it may be hard to understand. Empathy can also help the salesman save the prospect from future problems and maybe embarrassment as well. For example, let's say Mr. X is operating a small supply house out of a 5,000-square-foot building. He decides he wants to build a new facility with 20,000 square feet. He is ready to go, bubbling over with enthusiasm. He wants to push right ahead; but from what he has discussed, you put yourself in his place and ask yourself some very pointed questions—and come to the conclusion that Mr. X will be stretching things just to build 10,000 square feet. So you start working on the reasons why Mr. X should build only 10,000 square feet now and plan for future expansion. You may get a smaller order, but that's how good solid reputations are built, and reputation is very important to design-build construction business.

Empathy can also make your job easier. For example, you get the feeling a prospect is putting you off; he's not doing enough to push the project along even though both of you know he plans to build. Looking at the situation from the prospect's viewpoint may show you he really does have a time problem and is not pushing you around.

So far we have talked about the fundamentals that the construction salesman should be comfortable with: practice low profile, soft sell, and yet remain aggressive; think about how to push on buttons and how not to push off buttons; place yourself at the convenience of the prospect; learn to listen; impress the prospect with your interest in his business; accept the sometimes excessively long time intervals; be able to practice empathy.

All of these suggestions are only guidelines, not iron-bound rules. As I said earlier, selling is an art, not a science. Every individual who undertakes to sell construction projects aggressively is going to have to work out what does best for him. All I am saying is: here are some important areas that should be given careful consideration when you are eyeball to eyeball with Mr. X.

As you read on, I will refer you back to these basic guidelines with specific details to help you get a better grasp of them.

UNDERSTAND THE BUSINESS PROSPECT

At this point I would like to change the subject and consider the prospects whom the construction salesman will be calling on by business category. There are three main areas: retail, wholesale/industrial, and professional.

Retail. The first thing that you will notice about the retail segment of the business community is they think small (I don't mean this in any derogatory manner at all). The retail merchant for the most part sells items that cost from pennies up to several hundreds of dollars; he's used to dealing with small sums of money. At the same time his business outlook is shaped on a day-to-day basis.

Prove it to yourself. The next time you visit your local hardware store or drugstore, ask the owner or manager how business is. His answer will probably run along these lines, "Not too bad today, but yesterday was really off." To the construction man who thinks in terms of good and bad years, this day-to-day existence is a little difficult to fathom, and sometimes hard to deal with.

When the retail merchant has a bad day, then he has a hard time spending money; and believe me, one bad day can ruin the whole deal. For example, I was working with a retail merchant who had been leasing for years, and finally reached the point where he was ready to take the plunge and build his own place. Business was good, and after three months of preliminary work he told me to draw up the contract. The very next afternoon I stopped by for his signature, and was startled when he told me he wanted to hold off for awhile. The man sold toys, and that particular day was awful. Based on that day he decided to postpone the project until after he saw how his Christmas business would be. Well, I ask you, if a toy merchant can't look forward to and expect a good business during the Christmas season, just when does he expect to make it? The man never built; there was always some reason not to. I often wonder what would have been the outcome if I had had sense enough to visit him first thing in the morning before he had experienced that terrible day.

Timing is very important when you are calling on and working with the retail businessman, so make it a point to find out the best time to see him. Take the positive approach in this case—ask him. He will appreciate knowing you don't plan to take up his busy times.

It is maddening at times to try to carry on a meaningful conversation with the retail merchant when he's watching his sales area; and even if he uses his office or walks back to his stockroom, he is still subject to interruptions. Remember, as much as he may want to talk to you, his customers will always come first.

A method I have found to be most useful is to get Mr. X away from his place. The first way most would think about is lunch, and this is a fairly productive method. The problem with lunch is that there are always distractions buzzing around. The best way to get Mr. X away from his clerks and the phone is to take him to look at a building that is similar to what he has in mind. The prospect then becomes a captive audience, and the salesman can accomplish a great deal in a short time. Another good method is to arrange a meeting after hours, say in the evening or over the weekend.

The important point is that the retail merchant is at times difficult to talk with, and the good construction salesman is going to have to be able to maneuver the prospect into a meeting that will pay dividends.

Let's get back to the money angle. As already stated, most of the time the retail merchant deals in small sums of money, especially when compared to contractors. Say $200,000 to the average retail merchant, and notice the color slowly drain from his face and a slight glaze cover the eyes. Amounts of money that large literally scare the hell out of him. The thought of actually owing that kind of money is to him unthinkable, whereas the general contractor is used to running large sums of money through his books. The average contractor thinks no more of $200,000 than he does $2,000,000. The trick in the construction industry is to have some left after running these large amounts through the books. Right?

The retail merchant who is in the market for a new facility will generally have an idea of what the new building should cost. The figures will probably be at least ten years old. So when you give him some budget numbers, Mr. X will gasp and say, "I had no idea of spending that much; it's out of the question. I'll just stay put for a while."

At this point the construction salesman has to educate the prospect and slowly get him used to the numbers. This takes time and persistence. Remember, this money to the contractor is really not very much, but to the average retail merchant it's probably the largest money transaction he will ever undertake. He is reluctant and apprehensive. The salesman must understand this, accept the fact, and work toward making the prospect comfortable with the figures.

Sometimes the retail merchant does not have the financial background concerning large sums of money that you might expect him to have. He has no idea at all about the details of pulling together a financial package to pay for the new facility. Here the construction salesman can be most helpful and, at the same time, make himself more important to Mr. X—which is the name of the game at the selling stage. To sum up, the construction salesman must keep in mind when working with the retail businessman that this prospect is the exact opposite of the construction man in the manner that business is viewed and conducted.

Wholesale and Industrial. The second category, wholesale and industrial, is as different from retail as day is from night. This is the segment of the business community where you as a construction man will be able to establish the best rapport.

Like the contractor, the wholesale and industrial businessman is used to working with larger sums of money. This person does not freeze up when the salesman starts throwing around $200,000. Our prospect is right at home. The wholesale and industrial businessman does operate his firm on a daily basis as we all do, but he thinks in terms of good and bad months, quarters, and years. Also, within this time frame he can have seasonal peaks and valleys, like the contractor. Having to operate in this manner requires planning; therefore, our prospect will be more willing to listen to a good sales presentation and to appreciate ideas and help on future expansion.

Our prospect will have more employees than the retail man, including, in a great many cases, delivery personnel. He will have all the problems that go with having people on a payroll and trying to get the goods delivered to the customer.

The contractor faces the same situations trying to get his job done, and will have no trouble establishing a line of communication with this type of prospect, and through this, acceptance.

Let me give a personal illustration. In 1973 I was working with a good prospect who was in the wholesale food distribution business. I had already made two calls on this particular gentleman but felt that I was not really getting through to him. He seemed distant, and noncommittal. I was having trouble obtaining the necessary information to proceed with working up a presentation. I hung in there because the man was a first-class prospect.

The bank located next to his location had bought the prospect's warehouse after many months of negotiating. He had been given six months to move. Two months had already passed when I got into the picture, so Mr. X was very quickly running out of time, a factor that made him a prime prospect. During the second meeting I had determined that activity eased up during the middle of the morning, so I had a third meeting scheduled for the following Monday at 10 A.M. When I checked into my office, it was a typical Monday morning, and one of our truck drivers had stayed out, creating a bad problem on a special job. I had my secretary call, explain my problem, and push back my appointment. I finally solved the problem and showed up forty-five minutes late. The first thing Mr. X asked when I stepped into his office was if I had much trouble with drivers staying out on Monday mornings. It had been a hectic morning, and I just must have been looking for a sympathetic ear to bend. I took ten minutes, and told him how it was really difficult to plan for Monday mornings when you didn't know who was going to show up.

After my sad tale of woe, Mr. X sent for coffee and proceeded to tell me about his morning. He had two of his drivers fail to show. We spent the next half hour telling each other how hard it was to find good help this day and time, then the next hour working out the details of his new warehouse. He was a completely different person. Suddenly I wasn't an outsider pushing my way in, but a kindred spirit—one that could be talked to, and counted on to understand.

Two weeks later I had a signed contract for a 10,000-square-foot warehouse, with a 2,500-square-foot attached office.

The mechanics of dealing with the wholesale and industrial businessman are much easier than with the retailer. The former operates from an office behind a secretary. You will be able to make an appointment and conduct your business over a desk in an efficient manner. This helps the salesman because it means he will be able to put his time to better use than standing around waiting for the prospect's customer to clear out. It is easier for the salesman to drop by and hope to catch Mr. X in; and if he finds Mr. X in after all, the chances are that Mr. X will see him. Since he is not directly involved with the customer, our prospect has more control over his time.

After making several contacts, the construction salesman will come to understand and appreciate the wholesale and industrial prospect from still another point of view: *the size of the job.* These businessmen need space to carry on their enterprise: the wholesaler needs space to stock his goods, and the industrial man has to have space to manufacture his wares. These people think in terms of 10,000 or 50,000 square feet, and that's what makes the construction salesman sit up and take notice.

The important thing to keep in mind when establishing a contact which you hope will result in a contract is that the wholesale and industrial prospect is in a great many ways similar to the general contractor, and these similarities can be used. Better yet, they should be pursued and cultivated. The sooner a rapport is built, the easier the selling.

You may be thinking, why all the fuss over finding a common ground with the prospect? What's the big deal? Just talk football or golf, and do the *same job.* Try it, and you'll wonder what happened. The retailer has very little to nothing in common with a general contractor in a business way, so you maintain a pleasant, professional relationship and go from there. The wholesale and industrial people do have business areas that are very much like the general contractor's, and the good construction salesman will capitalize on them. But never try to make something up with sports or whatever. Realize this: these people you are wanting to do work for aren't dummies. They are astute members of the business community, and didn't get to where they are without some gray matter between the ears. They'll read right through a phony "hail fellow, well met" come on, their off button will start glowing a bright red, and you're dead. Even worse, they might tell their friends about you.

Now there are exceptions. Remember, selling is an art, so there will always be exceptions. If the prospect brings up the subject, okay, pursue it—but not into the ground; don't keep flogging it. It's little points like this, knowing how far to carry a subject for example, that make the difference between a good salesman and a super one.

Professional.   The third general category the construction salesman will be concerned with is the professional. I'm talking about doctors, lawyers, C.P.A.'s, engineers, dentists, architects, people of that caliber. Right away I'll say that the general contractor has even less in common in a business way with this group than he does with the retailer. These members of the business community don't have hungover drivers on Monday morning, and if there's a strike, they are just inconvenienced; the contractor can go bankrupt.

They seem to rely a great deal on other professionals when they need something done. When they are in the market for a building, be it a home or an office building, the architect is the first man they go to. There's absolutely nothing wrong with this. It is their choice, and I'm sure our professionals are more comfortable dealing with their opposite numbers. This does, however, have a direct bearing on the construction salesman because the job will most likely go out for bid if an architect is brought in at the very beginning. The bid job is the one thing the construction salesman should stay away from at all costs.

Other areas also cause problems when you are trying to deal with these professionals. They are hard to see! It's almost impossible to get past the secretary on the telephone, and don't even waste your time stopping by if you don't have an appointment.

When I phone, the type of secretary whom I dislike the most is the one who tells you in a crisp businesslike way, "Mr. X is very busy, but if you would tell me the nature of your business, I'll certainly see that it's brought to his attention." Usually the very busy fellow who is being guarded will be courteous enough to at least hear you out if you can only get to him. Well, in 1976 I ran into a secretary such as I have described above, and set my mind to getting through her, primarily because I had heard the day before from a friend that her boss, a lawyer, was thinking about building two warehouses (20,000 square feet each) for investment purposes. I must add that I felt there was a little bit of a challenge to try. I hung up, got a cup of coffee, closed my office door, and had a good think. I remembered the man's first name from the phone book; it was Edward. Saying to myself, "Nothing ventured, nothing gained," I called back, and just as Miss Ice Cube answered, I took charge: "This is Bill. Let me talk to Ed quick; it's about this weekend." Making the obvious deduction, she assumed it was an important personal call, and said, "Certainly, right away." Seconds later I had my man. In two minutes I had the information I needed. This Mr. X was part of a group, and I needed to talk with another party. He gave me the name and phone number, and he couldn't have been more pleasant. Oh! The group never did build the warehouses. But I do admit to enjoying getting around the guard; I still wonder what she thought if her boss ever mentioned the call.

When selling you have to play the percentages, and with respect to percentages the professional category doesn't offer as many business prospects as do the first two groups.

Okay, now for the exception: Professional people like doctors, dentists, and lawyers are constantly looking for places to invest money, whatever the reason. Commercial real estate can be made very attractive to them. The construction salesman can use this fact to go out and put together a venture with the clear understanding that he is going to build the facility. I must admit this is a special area where a thorough working knowledge of limited partnerships, real estate tax shelters, and mortgage money is necessary. If this segment of finance is new to you, then I suggest you make it a point to learn all you can because it will be a big help to selling. More will be said on this subject later in the book.

Let me summarize briefly what we have said so far. The methods that the construction salesman can use to gain entry to the prospect are very important. It takes two to sell something, and you are only half, so you must be able to confront and hold on to the other half. We have discussed what to do and what not to do to avoid turning off the prospect. The three general areas of the business community have been looked into, and their description will prove helpful in letting the salesman prepare himself mentally for the particular type of prospect he will be calling on.

Several more topics need to be discussed. They are just as important to selling as anything that has been gone over so far.

DON'T TALK DOWN TO OR PUT DOWN THE PROSPECT

This admonition will probably be obvious to the reader, but it still needs to be brought out and should always be in the back of the salesman's mind. The construction salesman calls on all kinds of people, and he has to be as adaptable as a foot soldier in the field. From the plush well-appointed office of a trucking company president who wants to spend $200,000 to enlarge his loading docks, the salesman may call on the owner of a small auto repair garage. No plush office with the attractive secretary bringing in the coffee here—the office consists of one desk in the corner covered over with paper of all kinds. It will most likely look like someone emptied the trash can on top. All the paper will be smudged with grease, and right in the middle will be two or three cast-off wiping rags. The couple of chairs around the desk can be classified as early greasepit. You hope above all that you can do your business standing up!

Now most garages these days will have a guard dog, and there he is under the desk giving you a beady-eyed look that says, "You're here only because I let you stay."

Mr. X comes up, and you introduce yourself and put out your hand. That's right, put the hand out; assume his is squeaky-clean. If he has a dirty hand, he'll most likely say so and let you know he'd like to shake your hand except for the dirt. But if he proceeds to shake hands, then grab it. When he points to a chair, sit in it; but as far as the dog is concerned, you are on your own—the best advice I can give is to ignore it.

Granted that this description may be a bit heavy-handed, I feel it makes an important point. If you look around as if you can hardly wait to get away, or don't sit, or maybe pull our your handkerchief to wipe your hands—then bang goes the off button.

My experience has been that the self-made man is a little more sensitive to the putdown, no matter how slight. I'm sure a psychologist can go on for days about why, but I really don't have the space for that. It must suffice for now that you be aware of this situation. This is another of those areas in selling that mostly has to be played by ear. I think the golden rule here would be an excellent guideline.

DON'T OVERLOOK SELLING TO THE PROSPECT'S EMPLOYEE

Whenever you get the chance, talk with the prospect's employees. It's surprising what you can learn that may be a big help with your negotiating. For example, I was trying to sell a motorcycle dealer a new building recently. He planned to relocate and had a new site; so he was a prime prospect. I kept going back time after time to try to get the project going so that I could put together a proposal. Well, after two weeks of feeling pushed around, I was ready to drop Mr. X and move on to something else. I decided to make one last call. I phoned, set up the appointment, and showed up on time to find Mr. X not back from lunch. Business was slow, so I struck up a conversation with the sales manager. We talked about a number of topics; then I brought the conversation around to the new building. The sales manager was very excited about the idea of a new building. He had a head full of ways to increase the business. I asked for his thoughts on several showroom and shop layouts that I was working on. The man was pleased as punch to help. Now, I honestly wanted his help, but at the same time I was selling myself to him. I wanted to find out if anything was happening that I should know about. I mentioned that Mr. X seemed unconcerned about moving along with the new building and was a hard man to do business with. Then the manager dropped the bomb. The building they were operating from had been condemned the previous month, and they had to be out in something like six months. Mr. X's attitude was just his way of trying to get the bottom price.

Needless to say, I stuck it out, and negotiations are still going on at this writing. Maybe I'll get the job yet. My point is: I was ready to give up until I fished about and made a big catch—which was exactly what I intended when I started talking to the sales manager. So chat with the hired hands every so often, and you may get some useful surprises.

PROSPECT EMBARRASSMENT

This area of discussion may be a little vague, and may even cause some outright laughter. But there is such a thing as prospect embarrassment. I know because I lost a nice $65,000 contract because of it.

Here is how it goes. When the salesman is working with a prospect from the beginning, there usually are many changes. This necessitates a constant flow of information from the prospect to the salesman. Well, after asking the salesman to change this and that for what seems like the tenth time, the prospect may begin to feel a little guilty about using the salesman's time. If the prospect is not 100% sure about going ahead, then he may just back off and lose his momentum rather than ask the salesman for something else. The salesman at this point may not be pursuing the prospect as actively as before for some reason or other. The project just drys up for lack of communication; the prospect is too embarrassed to ask, and the salesman does not follow up.

The next thing the salesman hears about is Mr. X building a new place, and he wonders what happened. Well, he has no one to blame but himself, and chances are that he won't be aware of what really went on. All the fault lies with the salesman for not making clear to the prospect that he is to call on him no matter how many changes are necessary. Make the prospect understand that it goes with your job, you expect to do it, and he should not worry about your time. Now the good construction salesman should be able to tell if the prospect is serious or is just manipulating him, looking for information that can be used later. This is a judgment call, and with a little time and experience to his credit, the construction salesman will learn to make the right call.

The salesman must *keep* impressing on the prospect the importance of always keeping the lines of communication open, no matter how much time is involved. Now the good salesman will not depend on Mr. X to do all the work. The other mistake the salesman must not make is failure to follow up. The prospect will not mind using your time so much if you offer it face to face.

I'm not going to say any more about prospect follow-up now. It will be covered in detail later on, but do remember that it is one of the most important aspects of closing a contract.

That job I lost was a warehouse for a heavy equipment dealer, and it seemed everything was always being changed. I started on two other projects when I was working with him, and I let things slip; and not hearing any more from the dealer, I thought he was still trying to decide what he was going to do. Two months later I ran into the dealer at lunch and asked him how things were coming on his building. He looked slightly uncomfortable and told me he had bought from another firm. Before I could say anything he went on, "I felt so bad about grinding up so much of your time, I didn't want to bother you."

I stood there dumbfounded! How could he bother me if there was a $65,000 contract involved? Well, I chalked it up to having to deal with individuals and the fact that everyone is different. Then six months later I was having coffee with a friend who builds custom houses, and performs a great many of the same services as a commercial salesman. We were comparing notes on selling construction, and he repeated a story that sounded very much like the one I have just described. He had experienced prospect embarrassment also. I have since talked with several other good construction salesmen, and they all agree that there is such a thing.

Now if you are skeptical, that's all right. But, please, when working with a prospect, tell him not to hesitate to call you. And you follow up.

STAND STRAIGHT AND LOOK THEM IN THE EYE

Do not feel that you are imposing on the prospect. Do not apologize for being there. Don't slip in the door and stand in the corner twisting your hat, hoping finally some nice person will speak to you in friendly tones. Walk in as if you know exactly what you are doing. *Look*, *be*, and *act* confident. No businessman respects a creampuff. And if you're wondering if you can do this and maintain the low-profile, soft-sell attitude, the answer is yes. It's not that easy to explain; it's something that just has to be done. I think the best advice is to practice the low profile, soft sell but give the impression that you are a businessperson conducting business during business hours and have no reason to be apologetic.

Matter of fact, I like to take the direct approach sometimes when I know something about the prospect. With the Mr. X who has the reputation of chewing people up who call on him, after doing what has to be done to get to Mr. X and while shaking hands, I look him in the eye and say: "Mr. X, I'm here because I can save you quite a lot of money on your new building. It'll take me five minutes to tell you how. Are you interested?" I have never had this direct approach fail. Mr. X always takes the time to listen. What that direct approach does is reach out and grab him by the shirt verbally. You have made Mr. X want to talk to you, and that's what selling is all about.

Now this direct approach will not be the best way all the time; usually it is best for the fast-moving fire eater type. The trick is to figure out what type of person you are calling on; so find out all you can about a prospect before you meet him. Calling a mutual friend is an excellent way. If possible, observe while waiting; doing so should give you a good handle. Another nonapologetic approach is one I use when the prospect gets into money and feels maybe the price is too high. I explain the cost of construction, and then, making it a point to look him in the eye, tell him profit is not a dirty word and that I fully intend to make a fair profit out of the job because that's the reason I work—the same reason that he, Mr. X, needs and wants the new building: to make money!

I generally don't use this approach unless I get the feeling that the prospect is looking down his nose at me; then I quickly yank him back to why we are talking and the fact we both are after the same end—making a buck! Never apologize for wanting to make a profit. It's really the main reason you are out there in the big bad world.

Remember, you are not only selling. You are operating in one of the truly specialized areas of selling. The construction salesman not only has to be a good salesman but has to have a good working knowledge of the construction industry. He has to be able to understand zoning, sewer and water problems, drainage, the various trades and how they dovetail together, architectural aesthetics, site work, paving, construction finances, permanent financing, and so on. In short, he has to be thoroughly knowledgeable about a very large and complex industry. At the same time, he is selling in the big league. Most salesmen close deals for $100 or $5,000; but the construction salesman closes contracts for a quarter of a million with the same ease that a tire salesman sells with for $80.

So hold your head up; few people can sell construction. It takes a special combination of knowledge, the hard hat in one hand and the calling card in the other.

I've only discussed the important areas of selling that apply to construction sales. Other segments have their special dos and don'ts. For example, the factory rep calling on his accounts on a regular schedule has certain rules he must follow, since even after he has opened the new account, he has to keep reselling every call. The real estate salesman showing houses may be really selling to the wife, which also requires special knowledge. There are many other aspects of the selling profession as a whole, but we are only interested in the significant information that will make for successful construction selling.

In closing this chapter, I want to point out that not everyone will be comfortable in this type of selling, and some will be comfortable only in certain parts—for example, the person who does not like cold calling may be a crackerjack when in his own conference room. But please take the time to give it a good shot. Who knows, you may uncover hidden talents and find that you really like selling construction.

Chapter 2

New Business Attitudes

You as a general contractor now have some idea what has to be done to meet the prospect and, just as important, keep the lines of communication open. When this communication starts to take place, the general contractor begins to find that there are new business attitudes he must seriously consider using because they have a direct bearing on successfully closing the contract with the prospect.

Most contractors feel comfortable in the bid market. They know the market, and it's only natural to be comfortable with what you know and understand. Along with this, I'm sure every general contractor who has been in business for any time at all has negotiated work. You are maybe beginning to wonder just what I'm coming to; after all, you've negotiated contracts, and they didn't require a bunch of new business ideas cranked into your operations. So what's the fuss, you're asking?

Okay, I'll answer your question. Ninety-nine times out of a hundred the negotiations you've made have been after you were low bidder and the owner wanted to cut the price even lower. Or you negotiated with an owner for a second project after completing the first job in a satisfactory manner. It could be a building that came from a friend or maybe the friend of a mutual friend of yours. In any case all these jobs have on thing in common: they were presold!

That's right; you negotiated the contracts *after* the owner was sold on your doing his work. There is one whale of a difference between that kind of negotiation and the ones in which you have to sell the owner on letting you do the work, and then negotiate the contract. The new business attitudes are necessary if you want to operate in this area of selling where the job has not been presold.

I would like to add here that your taking your time to read this book shows that you've already decided to look into and maybe use the first attitude on the list—which is to go out actively and ask for the business, to consider selling your services just like any other merchant.

I'm not talking about sweeping changes, but I am talking about spending some money. So please read this chapter with an open mind; don't form any iron-bound opinions until you finish the chapter, or, better yet, the book.

These ideas and guidelines are important to backing up the outside selling effort. I don't speak from theory, but from actual experience, the results of which have been nice sums of money in my pocket.

GOING OUT AND ASKING FOR THE BUSINESS

We have already backed into this business attitude so I'll not waste your time going over it again. It *is* a business attitude and as such must be listed here.

GIVING OUT INFORMATION

The general contractor is in business to make money. The one thing he does not do is give something away. Right? Not really, if he wants to sell jobs.

When bidding, the general contractor does not mind spending the time and money to put together a bid, and, believe me, I know how much of both will go into a bid. Then you, the contractor, go to the opening and sweat buckets. You're low bidder, and that's great. The only problem that nags at you is the $35,000 you left on the table.

The contractor goes through all these operations gladly because he's reasonably sure the project will be built. If the bids come in too high, then the low bidder still has a chance to negotiate the (presold) contract. So, all in all, bidding is worth the effort.

How would you like to spend some time working up prices, then give them to a prospect who doesn't have a clear idea if he is going to build? On top of that, he's not sure that what you priced is exactly what he wants. Can you handle this situation?

Now please remember what I asked a few paragraphs back, that you keep an open mind until you finish this chapter. Keep in mind the question I've just asked as we press on.

Bidding is not something a contractor thought up; it had to be devised by an owner. And who can blame the owner? He wants the lowest price he can get. The trouble with this system is that you, the general contractor, are always forced to tell the owner your lowest price. That's exactly why it's possible to leave $35,000 on the table; and I don't care who says anything to the contrary, it hurts to think you threw away a pile of cash. Then there's the bid you get that puts work on the books for the coming months, but very little profit. What's the percentage in it for you if you're just swapping dollars?

The shape of the bidding market is most likely the very reason that made you decide to give this book serious consideration. Okay. Now, remember the word empathy? (Look back if you're not sure.) The next thing we're going to do is empathize, to put you in the prospect's place.

You're standing behind your counter when in walks this fellow, who looks around and walks up to you. He doesn't look like a customer or like the type of salesman who calls on you. Your curiosity is aroused; who is he? The man puts out his hand, and introduces himself, then asks you if he can have a minute of your time.

Business is slow, so you tell him to go right ahead. You are still wondering what he wants. Then you find out. "Mr. X, I understand you are planning to open a new branch in the west end of town, and I'd like to talk to you about building the facility for you."

Well, he sure doesn't beat around the bush, you think.

"The best answer I can give you right now is yes and no. That's not much of an answer. You see, I have an option on the site, but I'm not sure I can swing it," you explain.

"Maybe I can help you decide," the construction salesman adds.

"I'd appreciate any help I can get right now. My real problem is getting some kind of handle on the cost of the building. Then there's the site work, and my real estate man has already told me there is a drainage problem that will have to be solved. I really can't make a decision until I get some idea of how much money will be needed. With that number then I can work out the projected business and know if I'll be able to cover the mortgage and over-head."

"How about giving me some of the details and letting me whip up some budgets for you to use?"

"Say, that'd be great; but wait a minute! What's this going to cost me?" you ask.

"Not a thing, Mr. X, except a fair chance to do the job. Now I'm not talking about a detailed quote, but a simple budget figure that'll not take much of my time. Really it's the only way to get the ball rolling. You need the figures to decide, and if you decide to build, then I have a good project to try to put to contract."

Let's look at our little scenario from another viewpoint, one in which you're still Mr. X.

There's no change in our make-believe situation until after you have told the construction salesman that you're not sure whether you will be able to build or not.

"Do you have any plans, Mr. X?"

"No, I don't," you answer, "but I do have some sketches," and you turn and hunt in your desk drawer until you find a line draw-ing done on a large brown wrapper. As you smooth out the paper on the counter top you think, maybe this man can get me on my way; I'm sure he'll be able to give me some ballpark figures.

"Is this all you have, Mr. X?"

"I'm afraid so. I didn't want to go to an architect until I was sure I could build."

"How about a site plan?"

"No." You feel that sinking feeling start; you know you'll get no help here.

"Mr. X. I'm sorry, but unless you have plans and specs, there's no way I can take the time to give you a price. I mean after all, you're not even sure you're going to build."

What are your reactions from Mr. X's viewpoint concerning these two interviews? The last one shows how most contractors probably would react; and you as Mr. X would write that salesman off and toss his card in the trash can. The first interview would offer you hope and a chance to get some information that could be very important to your pocketbook. You would keep the man's card and eagerly wait for the next few days to pass until you heard from him.

Okay, now stay in the guise of Mr. X, and we'll have the second interview. The salesman has called and set up an appointment at your convenience. You picked late afternoon because business is slow then. The man (who seems like a real nice fellow) shows up on time. After the amenities, he gets right down to business.

"Mr. X, the complete building shouldn't cost over $62,000. Now that is everything: plans, plumbing, lights, pegboard on the walls—in short, a turn-key job."

"That's a little more than I'd hoped for, Mr. Doe (you had sense enough to get his name off his card before he showed up). I was hoping for about $55,000."

"Well, there's a possibility it can be built for that figure, but please bear in mind this is a budget price and everything is estimated on the high side. I'm not helping you one bit in your decision if I give you a low price, and you're knocked over with increases after you are locked in. Best to be pleasantly surprised with lower prices from a not-to-exceed building cost."

"That makes sense, but how about the site?" you ask.

"I must say we had to hedge a little on the site work. The least will be $5,000 and the maximum can be $8,000."

"$3,000. That's a lot of difference," you point out.

"It sure is, Mr. X, but that bad drainage is the reason. There are various ways to solve the problem, and it will take some homework at city hall to see which way the city will allow us to go. [Note the *us* the salesman has just used. It's his first reference to getting you to thinking about the two of you putting this deal together. He's now started making you a little dependent on him.] Time is the problem, and I feel you should use the high figure added to your building cost to give you a good budget project cost. I would like to add that my firm has some cost-saving ideas that we hope will appeal to you if you decide to push on." (This is a good, low-profile sell; he's not pushing you; he's letting you set the pace, make the decisions to build, also offering help in the vital area of getting the cost down.)

Though the figures are higher than expected, you are pleasantly surprised at the interest shown. "This is just what I need. The numbers aren't completely out of line. Now all I have to do is convince the bank."

"I can give you some help there, Mr. X," the salesman tells you.

"How's that?" you ask, wondering.

"I'll give you a letter on company stationery outlining the project in general terms and stating the cost."

"Say, that'll be great. It'll make me look better down at the bank if they know my figures are from a contractor, and not snatched out of thin air."

As Mr. X, how would you respond to the two approaches? I'm sure you would be all fired up by the salesman who helped you start your project on its way. Suddenly there is somebody to light the way for you, to run interference; so there is an excellent chance you'll be able to build that new branch.

Let's stop pretending we're Mr. X and look at what our aggressive construction salesman did out of the ordinary. The answer is simple—he gave away information. A better word than "gave away" is "invested." Our man invested a couple of hours in time, and some of his expert knowledge in Mr. X in hopes that his investments would pay off in the form of a construction contract.

You may point out that at least you know the bid job will be constructed, whereas the negotiated job at the beginning may still be a toss-up, and that's the fact that does not appeal to you. I have two answers to this reaction. One, I firmly believe that the higher the risk in investing, the larger the returns, as money-lending firms point out. You can deposit your money in a savings and loan safely at 6½% or invest in real estate mortgages directly and make say 9½%; but you have a higher risk factor with the latter, hence more return. The same is true for construction selling; and, after all, is not more money per job the reason you want a piece of the negotiated construction market? The second reason is percentages. All contractors deal in the percentages of jobs bid against jobs contracted.

Selling is a percentage business also; so think about this: On a bidder's list of ten, you have a 10% chance to get a job that you are reasonably sure will be built. On a negotiated prospect where our Mr. X is not certain he will build, you know the odds are 50% he *will* build; and if you are the only salesman he is talking to, then you have a 50% chance of doing a job. Personally I prefer 50% to 10%. I know these percentages can change a hundred times depending on the circumstances, but, nevertheless, the contract-closing percentages for negotiated jobs are not as low as you may imagine. At the same time, if the negotiated job is going to be constructed, the number of contractors will be more like two or three trying to sell it instead of eight or ten as on the bid market. Still better odds in my book!

Providing information is one of the most important services a contractor can supply a prospect when the parties have a sell-and-buy relationship. It's strictly an investment and should be recognized as such by the general contractor, just as the time and effort to put together a bid represent an investment.

This investment will take the form of a budget estimate in most cases. It does not have to be a detailed quote, but only a ball-park figure based on your knowledge of construction costs. It's not too difficult to use square foot prices for certain types of commercial buildings.

Usually a simple line drawing showing the basic requirements of the prospect is the starting point. When meeting with the prospect, the salesman should know how to dig out information (details on this subject will be covered in later chapters) and put everything on the line drawing. This drawing doesn't have to be elaborate. I prefer to use $1/8''$ grid in a handy size, say $12'' \times 20''$. This allows me to have everything to scale, which makes for easier and more accurate budgets. There is also a small bonus, I feel—it looks good for the prospect to see you using professional material. Remember, you sell your competency every chance you have; never let up on this point.

Not only does the contractor have to accept this new attitude of investing his time and money in the prospect, but, along with this attitude, he must realize that the budget estimate he prepares will probably be the first of several. The first price is only the starting point. The prospect, upon receiving the first estimate, then begins to take an objective look at what he's trying to do. That's where the changes start, and these changes will require price changes. So be prepared to look at a prospect's job more than once; and, frankly, this helps the salesman with his selling. The more you can see the prospect, the more chance you have to sell him on yourself and your company.

I know that some skeptics are glad to hear that there is a limit to the amount of money and time that should be invested in the prospect. There will be prospects who keep changing; they never have any idea what they want, and will not hesitate to grind up your time for all the information they can obtain. (Incidentally, you don't have to worry about this type becoming embarrassed about using your time and services!)

It's difficult to tell at the beginning of negotiations exactly what type of businessman you are dealing with, so I've developed a rough guideline that does a good job without letting you get in too deep with a prospect who is just using you: After determining the requirement, I work up the estimate; and then I will make two changes if they are minor and require very little time. I only make one major revision after the first one. In most cases this will provide the prospect with all the necessary information to determine if he will build; or, if that has already been decided, then what he plans to build.

Please don't take this as a contradiction of what I mentioned in the previous chapter about prospect embarrassment, and telling the prospect to call on you at any time for changes. The pricing guidelines I've just described are guidelines, nothing more, that I try to use when actually putting out time and money for prices. I've found that the changes made by the prospect can easily be added to or subtracted from these figures. This allows the prospect to make changes, and I'm not really having to reprice the entire job.

The next step after running through the price exercises is for the prospect to make a commitment to you to do the project so that the final details can be worked out and priced.

There will be exceptions, and it's purely a judgment call on the salesman's part. Don't hesitate to do whatever is necessary within reason to provide the prospect with what he needs if you feel he is making an honest effort and is not just manipulating you.

There is another segment of the budget estimate to discuss, but before that I want to remind you that you're supposed to keep an open mind. Okay, here goes. It's pretty hard to talk to Mr. X about his new project without some sort of drawing. I know what some of you are thinking. First, I tell you to give away prices, and now I add plans! But before you toss the book on the desk, let me explain.

When you price a job to bid, an architect has already done all the work with the owner. It's been put on paper and given to the contractor to bid. With the design-build job the contractor is doing the preliminary work, some before the contract, with the bulk of it after the contract is signed. The one big difference for the prospect is that he has a true handle on the price when he deals with the contractor on a negotiated basis. At the same time, Mr. X cannot be expected to commit from price alone. Would you? I know I wouldn't. I would want to look at something!

The type of plan I'm talking about is a simple line drawing floor plan, and maybe a front elevation if it's a retail store, to show how the front will look. That's all I'm suggesting you use—not full plans and specifications prepared by an architect.

I personally draw my own drawings, and I'm sure every contractor can or has somebody who can make line drawings. These can also be a good barometer to see just how interested the prospect is. This point will be covered in detail in Chapter 4. I copy my line drawings on the office copy machine, and that's all I use.

I know for a fact that some of my competitors go so far as to have the drawings blueprinted. I'm told they feel this looks better, and I'm sure they're right; but I think that's going a little too far until you've been told you have the job. Also most contractors don't have a blueprint machine, so sending them out can run into money over a period of time.

It's difficult to have up-to-the-minute prices in all the various trades areas; also certain types of businesses will have special requirements that should be estimated by the particular trade involved. No matter what the reason is, whether price only or expert knowledge, trades will have to be called in to help with the budget estimate. It's important to have a working relationship with subcontractors who are willing to invest their time and money in your prospect. Their position is even less secure than the contractor's because the sub is not dealing directly with Mr. X.

My experience has been that the mechanical trades (heating, air conditioning, and plumbing) are the ones I always touch base with when working up a budget. Many times the sub will have to rough up a preliminary design in order to arrive at his price; so the line drawings that are used to sell the prospect also become the basis for the sub to budget his work.

There's nothing wrong with asking for prices from more than one sub in the same trade. The work may be complicated, and you need a good handle yourself before you give Mr. X his. The first budget from a sub may be too high; to use his price would raise your price higher than you feel will sell, so you look around for other prices in that particular trade.

I feel very strongly about the honesty and ethics involved in dealing with a subcontractor on a design-build project. If you use his price and sign up the prospect, then the sub should get the contract to do the sub work. I don't believe in trying to shop more profit into the job by putting that sub's work on the market; it's not fair to the sub because his budget price will have some money for the unknown hooker that can burn him—just as you should have in the budget you pass out to the prospect. It's very easy to shop around after the job is signed up and architectural plans are prepared because then the sub knows exactly what to price. I'm not saying for you to let the sub's budget price become his contract price to you; the sub will have to prepare a contract price from the final plans, and in most cases it should be below his budget because there are definite plans to work from. But he should be the only one pricing from the final plans, since he should be locked in to you as you are locked in to the prospect.

In the case where you know what you're doing, ball-park a trade price, and sell the job, then you can obtain sub prices to suit yourself; but please remember, if you use a person's time and money, then you should try to work with him.

Besides it makes good business sense not to whipsaw subcontractors about. They can be a good source of leads. This was brought home to me recently by an electrical contractor who had done work for me. He and his wife saw my wife and me one evening in a local restaurant, and came over to talk. While our wives were chatting, my friend told me about a lead. One month later I had the contract to build a 14,000-square-foot building. Needless to say, he did the electrical work.

I touched briefly on the final contract price while discussing the subcontractor, and I would like to elaborate a little on the subject. The budget price is used to sell the job; it's a tool, nothing more. It's supposed to enable the prospect to decide that you will be his contractor and also whether he will build. Now after this is accomplished, you, the contractor, have to get busy and have working drawings prepared in order to estimate the job correctly and present Mr. X with the contract.

Most of the time these will be the steps followed; requirements, budget estimate, decision by prospect, architectural plans, contract price. Once in a while the budget and contract prices will be the same, if the job is on the small side and the contractor has up-to-date prices he can use. My experience has been that this happens mostly with metal pre-engineered buildings under 3,000 square feet with very little or no interior work. With anything else I never take the slightest chance—prices are checked!

There will be details given about the budget estimate in Chapter 5, telling how to use it to best advantage when working with the prospect; but for now we will press on, since we are mainly concerned with the overall picture and how it relates to new business attitudes.

USE OF ADVERTISING

Let's go on to our next subject. Stay with me, keep that open mind, and I promise nothing will be as bad as giving away information and drawings. By going out to sell construction, the general contractor is now a merchant with wares to sell. He really is no different from the car dealer or beer distributor. You are going to have to start looking at advertising in a new light.

Let me say right now, I'm not talking about the special for the week as seen in the classified section of the local newspaper.

Job Signs.   Take job signs, for instance. Where else do you get the opportunity to put up a sign on the side of the road and tell the world what you do? Yet it seems to me that only the large projects get a decent job sign, and usually everyone who is anyone connected with the job has his name on it. What about the average job, say a 5,000-square-foot retail store? What's out front? From my observations it will be a two-foot by three-foot sign that shows its age. A prospect has to get out of his car, jump a ditch, and push aside some weeds to read the phone number. Well, I'm willing to bet that few do.

Granted that many general contractors are fully aware of company exposure from job signs and put up nice large four by eight ones. That's great, but how well are they taken care of? They look fine at first; then the weather makes them show their age, or maybe one of our Monday morning truck drivers backs into your sign, and it goes the rest of the job looking like a drunk leaning on a lamp post. Job signs are the best free advertisement you have; so it's important to use a large, well-done sign, and then keep it looking good throughout the entire job—and don't take it down until the very last minute. I've had owners ask me in a joking but partly serious way when I was going to remove my sign! So use them: big, well-done, and looking good.

Along with job signs go construction trailers, and you may use the side for your job sign. This is fine as long as the sign looks up-to-date. However, I prefer the job sign because it stays longer. You can put it up long before you move the trailer onto the site and leave it after the trailer is gone.

In Chapter 8 I will present details of working with the owner during construction, but here is the very first step, which I'm bringing up now because it goes hand in glove with job signs. When you price the job, add a few dollars to have an owner sign painted and mounted right alongside yours. The first weekend after signs are up, our owner will probably spend all day Sunday riding by to see:

New Home
of
Mr. X's Tire Service

You've put him right out in front for all to see. Mr. X is moving up, and he'll love it.

While we're on this subject, think about the location. Start out with the sign in the most advantageous place, even if it has to be moved later. A beautiful sign is worthless if not seen.

I know it's questionable to consider use of job signs as a new business attitude; but when some new thinking should take place about their use and looks, then I feel it should be included in this chapter.

Yellow Pages. The next pertinent subject is using the Yellow Pages. In my area only the metal pre-engineered building contractors advertise with more than their company name listed under "General Contractors." Unless you tell the public the services you offer, they'll think the obvious: that you do only bid work.

Take a good look in your Yellow Pages, and check more than one section. For example, under "Buildings" the metal building dealers all have blocks advertising their building name and the services that they can offer the prospect. Let me add that if you are seriously thinking of selling construction or have already started and don't have a metal building line, then you should give it thoughtful attention. I'm not preaching for the metal building industry; it's good business because they give you something else to sell, and the versatility of the metal building as well as the economics makes it a good item to use when negotiating the design-build commercial project.

People use the Yellow Pages. It's one form of advertising that the contractor will be able to monitor by logging the incoming calls. There is another big plus that is covered in detail in Chapter 4: the contractor will know the prospect is at least interested enough to call; a prospect who has already shown his interest is an excellent prospect.

So use the Yellow Pages and just don't list your name along with everyone else in the business. Spend the extra to have a block space telling the prospect what you offer. I think you'll be pleasantly surprised.

Newspaper Advertising. Ever thought about advertising in the newspaper? I know what you're thinking—another fool notion; but let's take a look. I know general contractors advertise in the papers because I've read the ads.

Metal building dealers run ads telling how great and wonderful their such and such kind of building is. I have reservations about the results obtained from this form of advertising. The odds aren't good enough for me. In selling you have to go with the percentages, and the odds on a prospect in the market for a new building looking at the paper and seeing the ad are just not that good. The same goes for TV ads. I've even seen them done, but believe me, not for very long, with TV time costing what it does.

I'm bringing up this form of advertising because I'm sure you will think about it in the future, and I want you to give it some careful thought before you spend your money.

There is one particular type of newspaper ad that I think does an excellent job of spreading the word. This ad is maybe a quarter page or even larger, congratulating Mr. X on opening his new facility. Usually the general contractor can get the subs and suppliers to chip in, and in return their company name also appears in the ad. Maybe some of you have already been involved with this type of advertising. The owner is pleased as punch to see the splash in the paper, and it is good business because a satisfied owner can be your best sales asset at certain times.

Admittedly, it's difficult to determine just how successful this kind of ad is. It's a lot like trying to assess the worth of job signs. You know they help, but you're not sure how much.

So take a look at your advertising. Bring in some professional help if you feel you need it. Just remember, when you have something to sell, spread the word.

ADVISER CAPACITY

The most amazing fact about prospects is their ignorance of the financial side of commerce outside their particular field.

When you are bidding a job, the plans are provided for you to use. You assume, most of the time correctly, that the money has been lined up and all the owner needs is a price and he's ready to go. In other words, most all the messy little details are handled for you.

Design-build is the opposite; you are going to have to help advise the prospect through these areas. Details concerning construction loans based on permanent money take-outs will have to be explained in detail.

You will have to make sure the prospect looks good when he approaches the bank. I don't mean a clean shirt, though I have worked with some prospects where it wouldn't have hurt. I'm talking about a package of information to be given to the bank people that tells them Mr. X has his act together and is providing them with contractor-backed information about the new facility he wants to build. Believe me, nothing does the job like a drawing when those conservative bankers are looking over the prospect's application.

I provide what I like to call a bank package for the prospect to present to the loan people when the job has progressed to the point where Mr. X wants me to do the work and will sign the contract contingent upon obtaining his mortgage. At this time during the negotiations, it's in your best interest to do everything possible to help the prospect secure the money.

My bank package consists of two to three typed pages outlining the scope of the project as well as general specifications plus a floor plan and elevations. I have the drawings prepared by a draftsman and then blueprinted. I have three, four, or five copies put together in a decent-looking folder for the prospect to pass out at the banks. I've found out it helps to have several copies for the convenience of the loan people at their meetings.

Along with advising the prospect on money, you'll be called on to answer questions about zoning, setbacks, sewer tap fees, and many, many more. It's possible the prospect cannot build what he wants where he wants, and you'll have to explain why and then help with the alternatives.

Now before you say that you just don't have the time for all this sort of thing, think about how Mr. X is becoming more and more dependent on you. This is an essential part of construction sales—to make the prospect dependent on the salesman.

As I have said, many businessmen are ignorant of money matters outside of their special field. This makes you a natural to be called on if the proposed project has trouble and the prospect cannot, or does not, want to own the facility but nevertheless needs the new building to operate from. In my experience almost any time the prospect has discussed leasing, it's been due to his needing the building and his capital also. Working out a lease package on a design-build situation can be very profitable to the general contractor, for if he has the means to obtain the site, either from the prospect who might already own it or by buying the site outright from a third party, the contractor will get a building contract plus the added value of investment property.

Most prospects really don't like the idea of becoming tenants and want to end up owning the facility at some time in the future; so they will be reluctant to enter into a lease agreement unless they have a buy-back option down the road, say no more than five years. If Mr. X wanted to rent all the time, he wouldn't be talking to you, but to a real estate agent with commercial property to rent.

Frankly, a lease design-build project is one in which everyone wins—and that's a good deal by anybody's yardstick. The prospect gets the building as well as having it designed to suit him; he holds on to his capital to operate with while his business grows, and then when he's able, can purchase the building. On the other hand, the contractor who signs a contract has invested in real estate that should have some tax advantages; and with what real estate has been doing these past few years, he should make a decent profit when he sells to Mr. X, since he sells at the fair market price at the time of sale.

Here's the hooker; nothing will grind up your time and money like working out the details for a lease, buy-back, design-build project—because you are doing it all! You're in charge of securing the construction loan, seeking permanent mortgage money, and handling design details with your new tenant, plans to banks, and all the dozens of problems that go with the building permits. When you roll the whole ball together, you'll wonder how it's possible to get it all done. But if you're willing to invest, then your return should be very attractive.

Selling construction will require that you act as an adviser from time to time, but at the same time it helps make the prospect dependent on you, so don't hesitate to do it.

I've made the assumption that the person going out to sell construction is knowledgeable in commercial real estate to the point where he can talk to the prospect about it. If you're not comfortable in this subject, then I suggest you become so. If not, you'll find yourself missing a real money-making job ever so often. Also it doesn't hurt to refresh the old mind in the subjects that might come up, such as setbacks, zoning, tap fees, whether gas is available, and so on. Make it a point to stay abreast of commercial money interest rates and general developments in the construction money area. Competency in all these subjects makes you, the salesman, look very good, and that's an integral part of selling.

REMAINING OBJECTIVE

I think objectivity is fairly easy to manage when the job has been bid off the exchange. Everything is cut and dried, spelled out right from the beginning. (Okay, I know that's the way it's supposed to be, but it isn't every time.) The general contractor has a long-range association with the owner, and usually the architect is also in the picture. This long-range relation allows the general contractor to keep everything pertaining to the project on a businesslike and objective plane.

Negotiated work is the opposite; you are dealing with an individual before the contract, not after. This means the contractor is vulnerable at this stage. He's putting out time and money with no guarantee of a contract. All you have to go on are the other person's words and your own judgment.

Let me give you a very good example: Several years ago I was working for a metal building contractor as a construction salesman. The owner passed a lead on to me concerning a certain Mr. X who planned to build a new automobile dealership. Needless to say, I went after the prospect hot and heavy. The car dealer had already hired an architect, and put out price feelers using his plans. He was just testing the market, he told me, before he went out to bid. I can appreciate this; he wanted to get a budget to use when he talked to the bank. He was checking to see if he would be able to build.

The tentative prices he got rocked him back on his heels. No way could he get that kind of money, he told me. He was in a holding pattern. He needed some help to move the project along, and I went to work selling. I sold myself, then my company, and went into saving him money to bring the construction costs down to near his budget, which was based on his borrowing power.

I put a great deal of time into the project. Fortunately we had the plans to use so my company wasn't spending money directly; we were spending it on my time and personnel in the office, however. After two weeks of work and three meetings with Mr. X, I called and recommended that the owner go back to the architect and have some large revisions made which would offer savings from the construction standpoint. I told Mr. X I could meet his budget with the new changes.

Mr. X was delighted with the proposal and wanted a meeting as soon as possible to get the project underway. This conversation took place on a Wednesday, and I wanted to have the meeting with Mr. X the next day, but he was planning to be out of town Thursday and Friday. He set the date for the following Monday afternoon. (I didn't like the long time interval for reasons I'll elaborate on in Chapter 5.)

I showed up on time for what I thought was going to be a sign-ing-up meeting. It didn't work out that way; it seems Mr. X had run into a contractor friend over the weekend who wanted to give him a proposal also. Being a friend, Mr. X agreed, and in good faith he couldn't sign with us and let his friend spin his wheels. He had no choice but to wait until the other price came in, which would be in two weeks. I left; there was absolutely nothing else that I could do. I stayed in touch with Mr. X for the next two weeks, and when the time period was up, Mr. X informed me that his architect had advised him to put the revised plans on the bid market.

I didn't like the turn of events one bit. Mr. X did assure me he wanted me to bid on the revised job. Several days later I picked up the plans, and when I unrolled them in the office, it was immedi-ately apparent that the price was going to come in way over bud-get. It was going to be a waste of time; there was no way that Mr. X was going to spend the kind of money his revised plans repre-sented. I decided not to do any more on the job. I was upset over the turn of events. I was going to sit back and wait.

Well, Mr. X was back to square one when the bid prices came in, just as I had predicted—way, way over budget. I honestly feel he was a little embarrassed to come back to me (remember what was discussed about prospect embarrassment); plus I think he was put out that my company hadn't bothered to bid. There was an excellent chance I could have salvaged the contract because I learned later that Mr. X had signed with a construction company that operated very much like my firm did, for a design-build job.

Do you know why I missed that job? I was ticked off and wouldn't follow up. I didn't remain objective.

It's important when selling construction to remain objective about the prospect. Don't let little things get under your skin. I repeat, you're dealing with the prospect one on one. His way of doing business may be different from yours; so don't blow what may be a super job because of emotional reactions.

I'm definitely not saying that you should let yourself be used as a doormat. There will be times when your decision will rightly be "To hell with it." I'm simply saying that before you take that atti-tude, you should give the situation a good think-through and make your decision based on hard facts, not because you're angry.

Recently I made the choice to back away from a prospect. This particular Mr. X was interested in adding on to his business; so I called on him. He was the hardest person to run down I've ever worked with. He had several businesses in the area, and no one ever seemed to know where he was. I chased him for two weeks before I was able to see him. That happened at a lucky stop at one of his locations. We had a good meeting, and I felt there were good possibilities for a contract, though one somewhat on the small side.

I put together my prices and called back when I was asked to. Well, the same runaround started again, and after one day of trying to find Mr. X, I said, "To hell with it." My decision was based on the fact I have other jobs to pursue, and I have only so much time. I didn't become upset with Mr. X because he's a busy man and very successful. I looked at the situation from the dollars-and-cents angle only, and came to the conclusion that I couldn't keep my schedule in disarray so that I could make contact with Mr. X whenever I got lucky.

I'll be glad to meet with him at any time at my office if he should call me back. That way I can work while waiting.

I'm the first to recognize that anyone in business will have unpleasant situations come up from time to time, and it's impossible to keep an objective attitude all the time. Okay, go ahead; make an emotional decision. I'll admit I've done so myself, but please don't make it a practice in construction selling.

If you got the impression that I've been preaching these last couple of pages, then I apologize. Remaining objective is crucial to selling construction; and, besides, I feel it never hurts to be reminded of the obvious every now and then.

I'll leave you with one parting shot before we move on to the next chapter: Remain objective about what has been gone over in these first two chapters. Even if not sold on an idea, consider giving it a try, but with an open mind that will let you remain objective. At the same time be flexible and use what you find works for you. The first two chapters really are general in nature and give the broad foundation that launches selling construction.

From here on we'll be dealing in nitty-gritty details; and time after time, as these details are explained, we'll be referring back to topics that were covered in general in these first two chapters.

Chapter 3

Finding the Prospective Customer

About now you're thinking, what good are all these guidelines, theories, and ideas going to do if there are no prospects to sell to?

This chapter will solve that problem for you. I'm sure everyone has heard the old adage "Gold is where you find it"; the same applies to construction prospects. If there is one certain phase of construction sales where plain old hard work pays off every time, it's "digging up" the prospects. It's simply percentages—talking to enough people, looking deep enough in the right places, and keeping your eyes and ears open will always pay off, and I mean *always.*

Throughout Chapters 1 and 2 the word "prospect" was used to denote the businessman who was in the market for a new building. At this point I want to bring in a new term to call our Mr. X before he becomes an honest-to-goodness prospect: Mr. X is a *lead* before he becomes a prospect.

All of you have watched enough TV police programs to be familiar with the term lead, and with how leads are used. Well, the good construction salesman will use leads just as the good police officer does. A lead can be a name or something you saw at someone's business address or a phone call from the Yellow Pages; leads can come from anywhere. They are the one thing you cannot do without—no leads, no prospects, no contracts, no money. It's that simple. Until you qualify the lead and confirm that he is in the market for a new building or some kind of construction work, he's not a prospect.

Now the main thrust of this chapter is to show you how to locate the leads that can be turned into prospects. So please don't confuse the terms lead and prospect. To reiterate: A lead is someone you think might need a new facility. A prospect is someone you *know* is in the market for a new building.

It's important for the construction salesman to have a prospect list and keep it constantly up-to-date. To have a productive prospect list, the salesman has to check out and follow up on a large number of leads. Again it's percentages; many leads make a small prospect list that boils down to one or two contracts. Obtaining and checking out leads is one of the most important functions the construction salesman can perform—or any other salesman, for that matter, regardless of what is sold.

This chapter is about how to go about obtaining those all-important leads; so let's get right into it.

PERSONAL CONTACTS

This subject can cover a great variety of situations. You're at a party and a friend comes up and tells you about someone who's planning to build. Or it can be your banker or a salesman calling on you who passes the word.

Some of the personal-contact leads are just plain luck. That friend you saw at the party might not have taken the trouble to call you at your office. There's nothing you can do about the leads you miss because somebody fails to tell you, but there is one positive action you can take—make sure everyone you associate with knows what you do. If as a contractor you're going into the negotiated design-build field, then let your friends know, both social and in the business community.

There is a courtesy I always observe when a lead has been passed on to me personally. After I've checked the lead out, I contact the person who gave it to me and give him a status report. The report may be, sorry Mr. X isn't going to build, or he's thinking about it, but his lease has one more year to go. The important fact is you show the person you report to that you took him seriously and you appreciate being told. At the same time your friend has the straight scoop about someone else in business, which might make him look good by being knowledgeable. This sometimes can be most important, since everybody likes to have inside information.

A fine source of personal leads can be clubs, either social or civic. If you don't belong to any, then I suggest you look into some. It's very important to know the right kind of people in the business community—the bankers, real estate brokers, lawyers, and loan company officials. They are the people located at or near the money, and everything in business is either going to or coming from the source of money.

Remember, it's strictly percentages; the more people you know and who know what you do, then the more leads. To be perfectly realistic about it, the old saying "It's not what you know but who you know" is very accurate. For example, two weeks ago I closed a contract that came from a neighbor. He told me that his employer had to move, his building had been purchased by the government, and why didn't I talk with him.

Right now I'm in the process of negotiating with two other parties for new buildings, and both men belong to the same club I do. I was approached during lunch at separate times and asked to stop by when convenient. These three leads, and that's all they were until I checked them out, came from personal social contacts, and everyone involved knew exactly what services I offered.

So get the word around town; it'll pay off for you.

There are two particular groups in which I make it a point to have personal contacts, the first of which is commercial real estate brokers. Nothing will happen until Mr. X has a site; so the broker can be a good source of leads. At the same time, the broker may need your help because Mr. X won't buy the land until he has some idea of what the entire project will cost, and you'll be able to provide those numbers right up front. Develop a rapport in the commercial real estate business, and let the brokers know you stand ready to help them close a deal.

It's really a two-way street, and the contractor who's actively out hunting for business will turn over leads who are thinking about building but don't have their site. Thus you are able to pass a lead on to a commercial broker. Now you're helping yourself, since when Mr. X obtains his site, he becomes a prime prospect.

There is one drawback when you are passing the lead, however; this is security (to be covered in detail in Chapter 5). Unless the broker can keep his mouth shut, don't use him. It's extremely important that the lead not get out on the street. If you dug the lead up, there's a good chance Mr. X may not be working with anyone else; and the fewer who know about Mr. X, the better your chances are for a contract.

So work with a real estate broker who will work with you in every sense of the word, and make money for the two of you.

The second group that is a good source of leads is the subcontractors you work with, mainly the heating, air conditioning, electrical, and plumbing companies, since they will be doing different types of commercial work for the business community which is not new construction. A great deal of their work can be repair, maintenance, and renovation. These people will be going into a company, and are in a position to notice and hear what plans the company may have. It can be such a simple remark. The heating man asks why the new ductwork is not being extended to the rear section of the building, and the answer is that that section is to be replaced in the near future. That's all it takes, and the sub should be on the phone to you passing on what is obviously an excellent lead.

Now if you have the proper relationship with the sub, he really is passing the lead to you mainly to get himself a sure job—he knows that you're pretty good at negotiating contracts and that if you get the job, then he has the sub work in his trade. In other words, he's looking out for number one, but in the process he lets· you do the same.

Now remember back in Chapter 1 we talked about aggressive selling. Well, there's more to aggressive selling than looking the prospect in the eye and telling him you want to do his work. You have to be aggressive in other areas, and that means taking the initiative. Don't sit around and wait for those people who can furnish leads from the business community to give you a call. People are generally busy and it's very easy for someone with the best intentions simply to forget. So you give them a call every so often and ask if they have anything that looks promising.

Learn to pick up information every chance you have, whenever you see people who can furnish leads. (I'm speaking of business friends now, not social. I feel everyone has his own manner in mixing business with pleasure. My personal feeling is that it's okay if the other party approaches you, and then you don't stay on the subject very long. If it's something that should be gone into in detail, then make an appointment for business hours, the same as any other professional would.) As I stated, learn to ferret out information from these business sources of leads. Ask some questions.

Let me illustrate. A real estate broker who has his office in the same building I use stopped me in the hall one morning and asked if I could give him a budget price. He was trying to put together a lease deal with a hardware company that wanted to expand. The cost of the construction was needed to work out what the lease amount would be.

Naturally I agreed and turned the price over three days later. Two weeks went by; then I ran into the broker in the parking lot and asked how the negotiations were progressing. He told me he'd lost out to another group who'd also been making a proposal. There was a time when I'd have told my broker friend I was very sorry to hear he had lost out and walked off. But I've since learned better. When he told me another group had the lease, I had a hot lead to be pursued immediately. I casually asked who was the other group. The answer was some lawyer, and he thought his name was this or maybe that.

I thanked him and went straight to my office and called my attorney. He gave me two names who he thought it might be, and on the first call I hit pay dirt. I went to work and presented my proposal several days later. The story doesn't have a happy ending, though; I didn't get the job. Nevertheless, the pertinent point is the manner in which the lead was "dug up."

Remember to dig (I know I've overworked that word, but it's the best word to convey the meaning I wish to put across); practice pulling out every scrap of information when there's a possible lead involved. People know more than they realize.

REFERRALS

Referrals are different from personal contacts, in that they come from people you've done work for in the past. A referral can also come from another contractor who can but does not want to work with the prospect for whatever reason. This is an extremely valuable source of business because you know the lead is interested in doing something if he goes to the trouble of talking to an owner and then contacting you. The problem with referrals is the time lag involved before they begin to pay dividends. You might wonder how this can be, since you have been contracting for years and have plenty of owners out there. But think: What kind of owners? Is it one who says, when asked by Mr. X about his contractor, "Good outfit, but you're going to need your plans first before they'll talk to you"? Or will Mr. X say, "Good outfit and they can take your project right from the start and do the whole thing for you, a complete turn-key job"?

In my opinion the only way to get the second recommendation is to be in that type of design-build work.

There's another point; most bid jobs have rather long-range owner relations, and when the problems begin to pop up, the contractor retreats back to the plans and specs and makes little effort to work with the owner. This has to color the owner's thinking about the contractor; so why should he pass out glowing reports about how great you are?

No, you're going to have to build your turn-key job reputation, and that takes time. It'll be a while before the referrals start coming in; but when they do, it'll be the source of leads with the highest percentage of contracts. Compared to the other sources, the number will be small; but the results can be outstanding.

ADVERTISING

I've already touched on this in Chapter 2, but let's look a little deeper.

Yellow Pages. I feel use of the Yellow Pages is one of your more successful advertising methods. I find myself turning to them all the time. It's possible to keep tabs on Yellow Pages results by the incoming calls, but you're going to have to instruct the person answering the phone to ask the caller nicely if his call originated with the phone book. Study the ads that your competition has and work up something for yourself. I'm not saying to go for a big splash, but I am saying to be in the Yellow Pages. I imagine your local phone company can be of some help here.

Now incoming phone inquiries are very good leads because the people are coming to you, which expresses interest on their part. At the same time, you or whoever talks to the incoming caller will be able to secure necessary information, and this will save the salesman time.

If your office is like most, you have someone who handles the phone, and often there'll be no one to talk details with the caller except for the secretary. In this case, train her to ask the proper questions so that the salesman can do some homework before he returns the call. Many times the lead is no lead at all, and a smart secretary can again save time.

It's important to find out exactly why the caller is calling. If you do only metal prefab buildings and the caller wants a block addition put on a shopping center, then your secretary should be able to handle the situation and hopefully refer the caller to another construction firm. If the caller turns out to be a bona fide lead, then your office help should try to determine the building size and just how the building will be used. The salesman can have his mind in gear when he calls back, and that's important in helping the lead start to develop a good impression.

A word here about calling back. DO IT! And promptly. That's the first good impression the caller will form. I'm absolutely amazed at the number of businessmen who don't take returning phone calls seriously; but for a construction salesman, it's just plain foolhardy not to.

Direct Mail. My experience with direct mail is that it's not very effective in providing good strong leads. Leads, yes, and you'll get everything from the high school student doing a paper on the building industry to the guy who is thinking about opening a small business when he retires in ten years and wants to start getting some ideas.

The metal building industry seems very big on direct mail campaigns, and the home improvement people make good use of this type of lead gathering. As far as the design-build contractor is concerned, in my opinion it's a waste of time and money.

Again that well-used word percentages: it's just like newspaper advertising—the odds are not good enough to hit the man who's going to build at the right time. Very little to nothing useful will come out of a direct mailing, so I recommend that this form of finding leads not be used.

Publications. I've already covered newspapers, and my feelings here are the same as for direct mailings, with the possible exception of the grand-opening type of spread, which is more happy-owner-oriented than suited to lead gathering.

I'm also of the opinion that radio and TV won't offer any better results than direct mailing and newspapers.

I'm sure you may think about magazines or booklets, both within and outside the construction industry. There's not much here either to bring in leads. Construction industry publications are a complete waste of money; you certainly aren't going to sell another building to some other contractor. The non-industry publications carry the same long-shot percentages as direct mailings, newspapers, radio, and TV.

My experience has been that Yellow Pages and job signs are the two best methods to obtain the caliber of leads that can be turned into contracts. In the advertising area this is where I would strongly recommend that you spend your money.

Company Brochure.   Before I leave the subject of advertising, I'd like to mention company brochures. Many construction firms have them, and I've seen some that were extremely well done and did a beautiful job of telling the company story. I'm bringing company brochures up here because they really are a kind of advertising and can be a very useful selling tool. However, they are of very little use in bringing in leads, and for such use shouldn't even be considered. I might add that a well-done brochure can be very expensive. It can still be well worth the money, though, as a selling tool, especially on the larger, more sophisticated projects. But if you're just getting into construction selling, I suggest you let the brochure come later.

The lead sources that have been discussed so far are matters of just good common business sense. Now I'd like to talk about some of the not so obvious areas in which to obtain leads.

TRENDS IN SPECIFIC BUSINESS AREAS

This subject will take a little more work and observation on the salesman's part. Business trends of all types are taking place constantly. Some are quite obvious, and then again some won't be noticed unless you take the trouble to look. For example, back in 1973–1975 when most of the nation was in what was politely referred to as a recession, the contractor and many other people found it hard going just to keep the doors open. Kindred souls were the automobile dealers. Can you remember back when they were doing anything to sell a car? While the car dealers were having rough times, as were many allied businesses in the automobile industry, guess what were growing and doing well? Auto parts stores. People were driving what they had longer; therefore, they needed parts. Also prospering were repair garages, which made money keeping the old car on the road, at the same time buying parts from the parts suppliers.

I know of one contractor who sold three auto parts stores in 1975, when other contractors were telling each other how bad things were. My point is that even in slow business cycles, somebody will be making money. All you have to do is pay attention, observe, and ask questions; then go after the people in that particular market.

Motorcycle sales have gone up astronomically over the past few years. There are many reasons why, which are not our concern; but the fact that sales are up makes cycle dealers a prime target for leads. At the same time you are doing your homework on a business trend, make it a point to learn what you can of a general nature about the business. Sometimes you can get help from an unknown source. We're talking about the cycle dealer; so consider this: I understand that most started out as second lines for car dealers, and the motorcycle manufacturers don't want their cycles marketed alongside cars—they want a completely separate facility for the bikes. Some dealers could be required to separate the two; in some cases it could be the car manufacturer asking for the split. Whichever, this kind of knowledge gives you a good source of leads that you probably wouldn't have pursued had you not taken the trouble to learn something of the industry you are interested in.

The trend can be going on in any business area; all the salesman has to do is dig around. Talk with people who deal with the whole spectrum of business—bankers, stock brokers, real estate brokers, lawyers, and mortgage loan officers. Talk with friends who are in other business fields. A good place to maintain good relations is city hall. Maybe there is some new regulation that will cause some special construction to take place. For example, you might discover that the health department has a new ruling about the storage of certain kinds of food, and the food distributor will become a prime area to be worked.

Trend leads do have the drawback of being somewhat long-range if the salesman is on the ball. That is, he may call on Mr. X before he is really ready to expand. Mr. X knows his business is good; and if you caught him early, it'll take some time before he's ready to build or add on. Once in a while you'll catch a fast-moving trend lead, but my experience has been that they have to be nursed along for a period of time. How long? I can't say. We're dealing with people; therefore there's no yardstick. I'll give it one quick swing based on past experience, though: six to twelve months.

I consider publications as a good source of trend information—the better news magazines, financial papers, and trade journals for other fields. I make it a habit to look through any trade magazines that are in the lead's waiting room, and you don't have to be in an office. Just about every business will have trade publications lying around somewhere. Look at them; never forget that leads are where you find them.

Trend leads require more bird-dogging on the salesman's part, but trends are nevertheless a very important source of leads and should not be ignored.

REDEVELOPMENT PROJECTS

Redevelopment projects can be a veritable gold mine of leads. It seems that almost every city has a program to buy up old sections of town, bulldoze the buildings, and rebuild. The businessmen in these areas are being forced to move, making them prime leads to be called on and worked. I have personally found that many of the older established distributor types of business are located in these older areas of town, giving the construction salesman not only an excellent lead but one which appeals from the standpoint of both job size and building an easy rapport with the buyer.

The place to start is at the local redevelopment project office. Gather all the pertinent information possible, talk with officials, look at maps, and come away with anything they'll give you. The redevelopment officials will certainly not be able to recommend you to a prospect, but they should have no qualms about telling you about Mr. X down on Sixth Street who's dragging his feet about leaving. The officials may be good enough to pass on lists of names whom you can contact, which allow you to call first. One very important fact should be uppermost in your mind, though; the lead is faced with having to move. No if, ands, or buts—he's going. There may be a time element involved, but at some point he's going. Rest assured that you cannot find a better lead to work.

Now if you are aware of redevelopment projects in your area and are familiar with them, you may feel there's no need to waste your time talking to the officials. You'll go right down and walk in and meet Mr. X. Don't do it; always check in with the office, and learn all you can. After all, the area you can see being redeveloped may be a small part of a much larger whole that only shows up on some map in the local office. While you are at it, make it a point to stop by every so often and keep up-to-date on the changes.

RELOCATION PROJECTS

Relocation projects are similar to redevelopment projects but have one main difference: when you see the work taking place, chances are you're too late to sell a job. The relocating businessman who has to move will have all the details behind him before he moves out and the bulldozers move in. Again these leads are prime ones because Mr. X is being forced to move. No question about it; he will build a building.

Relocation projects bring to mind right away new highways or roads; and you're right—this area offers very productive leads. At the same time, expand your thinking into other fields that might offer leads. How about business areas around a fast-expanding unit, say a college or a hospital complex? Military installations offer good leads, and I've just had this brought home to me in a very positive manner. The Langley Air Force Base enlarged the safety zone at the end of one of the main runways; and the government bought out three businessmen. Two rebuilt, and I negotiated a contract with one for a 2,600-square-foot office building. Point of interest: The other owner acted as his own general contractor and rebuilt, so I didn't lose him to my competition.

You have to work at finding relocation leads. Watch the paper; my experience has been that relocation projects are advertised in the papers as well as being carried as a news item. The second way to discover what's going on is to find out the government agencies (federal, state, and local) that are concerned with relocation and then go see them from time to time.

Of course you'll learn when you start looking for leads that one lead-gathering method will help with another; the lead will be the result of the combining of two or more methods. Say, for instance, a friend at a party happens to mention a relocation project—no names, just a lead for a lead, which provides the salesman a prime area to investigate.

I repeat: Leads from relocation projects are extremely productive because Mr. X has to move.

INDUSTRIAL COMMISSIONS

Most larger cities are very interested in bringing new industry to their locale; therefore, there should be an industrial commission that the construction salesman can contact for leads. Many times the commission operates an industrial park so that ready sites can be provided to the prospect shopping around for a new plant location.

It will pay the salesman to check out this area and become acquainted with the officials. I do want to point out one important fact: this type of lead is not as productive as the ones we have covered earlier, the reason being that the lead in the beginning is just looking around. He has no site and maybe is not even sure of the locality when he wants to build.

Work with these leads only after you know that a site has been obtained. The out-of-town lookers are notorious for taking your time and information and then moving on.

Don't ignore this source of leads, but keep both eyes open when you're working with this type.

REJECTED BIDS

This is one segment I particularly enjoy working with. I've had several good money-making jobs to come from the rejected-bid ranks. This source of leads is also an easy area to pursue. Just sit at your desk, and look through the material from your bidder's exchange, and then pick up the phone. Now rejected bids are not leads but prime prospects because the owner has everything together and is ready to go. Then, wham! He's knocked over by the prices and wonders what to do next.

Suddenly the construction salesman appears on the scene and starts showing Mr. X how to lower the cost and hopefully build the facility without losing needed requirements. Chances are, Mr. X has already talked with the architect in charge and learned the universal fact that most architects cut costs by cutting square footage, both in the building as well as on the site. It's perfectly natural for them to do so; they're professionals who have to go through college, then qualify to be licensed in the state in which they wish to practice. The problem, I believe, comes from the fact that they don't operate on a day-to-day basis with actual construction costs and how these costs relate to different forms of construction as well as materials.

A personal experience will illustrate exactly what I'm talking about: In 1973 a local equipment company had plans prepared for a new facility; it went out to bid and was floored with prices $50,000 to $60,000 over budget. I came across this fact through one of its outside salesmen before it became public knowledge and immediately contacted the owner. I picked up a set of plans and my company studied them for a week with the aim of cutting construction costs without decreasing square footage. At the end of the week I met with the owner and told him we could not only lower costs, but there was a good chance the final price could be under the original budget. The owner, who was one sharp individual, answered my presentation with two words: show me! I explained that if he could spare the time I wanted to arrange a conference at my office so that the estimator and owner of my company could explain in detail plus have the figures handy if he wanted to think about other changes.

My reasons for setting up the meeting were twofold: one, I did want our estimator, who was a licensed structural engineer, to explain in detail what was proposed; and second, it's a good selling technique to show the prospect he's dealing with a first-rate outfit—and what better way than to let him see a busy, well-run office? (If your office operation will not give this impression, then don't use this idea.) Okay, back to the meeting. We showed the owner that the structure he wanted to build was basically a metal building but not of the standard pre-engineered variety; also the exterior panels were of a very expensive material. Oh, no doubt they would look good, but could he afford that much money for looks? No was the answer.

Our recommendation was to substitute a standard pre-engineered metal building with the manufacturer's standard exterior panels. This represented a substantial savings. Another big savings came from the parts department. The original plans had had this area air-conditioned because the owner had so instructed. No one had bothered to explain to him that this was a large volume of space to cool, and it would be costly. We recommended instead using a mechanical air-change system that would bring in outside air while exhausting the warmer air from inside the area. After all, the shop wasn't air-conditioned, and the parts department was attached to the shop area. Why hand a nice cold part to a mechanic when you were paying to cool it? There did have to be some provision for the parts people to be able to work in warm weather; hence the mechanical air-change system. All the mechanics had to do was open the huge door at each end of the shop for extra ventilation.

Another place where we saved money was the glass in the front office section. I recall that a special bronze glass was specified, but the front of the building had an overhanging mansard. Also many large oak trees shaded the entire front. We suggested regular clear commercial glass, since the sun's heat wasn't a real problem.

We suggested quite a few more money-saving ideas; but they were all small, so I won't get into them now. Needless to say, the owner contracted with my company to build his new facility, incorporating the money-saving ideas. Our final price was indeed under his original budget figure, and he didn't lose one square foot of area. It was also a profitable project for both the company and me personally.

The ironic thing was that the lead didn't come from the bidder's exchange but from a personal contact, which goes to show that business is where you find it, or in some cases, where it finds you.

The point to remember about rejected-bid prospects is speed. You have to move fast. As I said earlier, most negotiated jobs take time, but the rejected bid is the exception. Our Mr. X is ready to go; everything has been worked out—site, plans, money in most cases; all he needs is some help getting the price back down to earth. Unless there's a special situation with a lead or prospect, I personally put the rejected bid right up front, with precedence over anything else I happen to be working on.

While on the subject of rejected bids, I made the assumption that the construction salesman or someone else in the company has the ability and know-how actually to redesign a commercial project, with savings recognized from different construction techniques and types of materials. You will not have much of a chance to sell the rejected-bid prospect if you can't offer him some tangible evidence of how you're going to do what you say you're going to do. If you need help in this area, get it; don't try to bluff your way with the prospect. At some point it will show that you don't know what you are doing, and Mr. X will give you your walking papers.

Not all rejected-bid negotiated contracts will go as smoothly as the example I have just described. Sometimes the owner won't be able to get all his requirements. It may not be feasible to bring the cost down to what the owner is looking for without some compromise on his part. When this happens, I push the idea of "better something than nothing." I stress compromise now with expansion designed into the plans to pick up the dropped requirements in the future. When informed of the cost of items in the original job plans, the owner may find there are some he can do without. For example, when Mr. X starts moving up in the world, he tells himself he's going to do it right. So he instructs the architect that he wants a large office with adjoining bath. Then, almost as an afterthought, he adds a shower to the bath. Mr. X is a busy man and many times would find it convenient to change clothes before attending a club meeting in the evening. The architect says, "Sure thing," and draws in a 20' X 20' office with a ceramic tile full bath.

Now if the above job reaches the rejected list, the construction salesman may have to ask Mr. X if that big office and bath are really necessary because they're costing so much money. Could he get along just as well with a 12' X 12' office with a half bath that doesn't have ceramic all over it? My experience has been that the owner quickly cuts out what are obviously luxury features and starts working with the hard monetary facts so that he can build the project he wants so badly.

The luxury compromise doesn't really cause the owner very much concern; he's a businessman and facts are facts when he's faced with the decision of whether the money spent will help him make money. On the other hand, a compromise that may interfere with his making money is something not to be taken lightly; it requires some thought.

The future owner may conclude that he's going to cut and compromise only so much. His next step will be to accept a higher budget and go after more money. Guess who will be standing right there with dependable, realistic prices plus any other help when Mr. X returns to his bankers? That's right—you!

The one thing the construction salesman must always remember about rejected bids is to get the job off the bidder's market and then have Mr. X commit to him as quickly as possible. Never lose sight of this; the longer the project stays on the street, the better are the chances that something will happen to foul you up. Believe me, it can be some small insignificant point that blows the whole deal.

Remember, the rejected bid is not a lead, but a prime prospect because all the preliminary work has been done. He wants to build, so give the prospect top priority and move fast!

COLD CALLS

In this particular type of lead gathering, the construction salesman must be able to think on his feet and observe details. Many salesmen don't enjoy making cold calls; they're uncomfortable going in when not expected. To be successful, the construction salesman should adopt the same attitude the old prospectors had, that "the next rock I turn over could have the mother lode under it." Every door you walk through could have a half-million-dollar contract waiting.

You may be asking yourself, what's the difference between cold calls and direct mailing? After all, it would seem that direct mail would be cheaper than paying a salesman or using my time to cold call. There is one tremendous difference. The postman delivering the mail doesn't look the lead over; he merely drops the mail and moves on. When a construction salesman make that cold call, he's been able to observe the physical surroundings. Maybe he noticed material stored in the open or obviously incoming boxes left on the sidewalks; or it could be a garage almost covered up with cars to be worked on. In other words, the salesman is able to observe that the businessman might be in the market for a building, and armed with this knowledge walk right in and introduce himself.

Remember also what happens to direct mail once it arrives at the lead's place of business. You hope someone just looks at it; and for someone to read it is beyond your wildest dreams. A person standing in the office is not so easily dismissed; he will under the worst of conditions at least learn the name of the person he has to talk with, and can then follow up by phone. Believe me, the postman won't do this for you.

In my opinion there's no way direct mail can take the place of the cold call. I'm talking about direct mail in the true sense of the definition, which is mailing to an address blindly hoping something will come of it. Please don't confuse direct mail with follow-up mail. Follow-up mail is a very useful tool for the construction salesman. Hopefully it will put his name in front of the lead, or maybe the letter can be used to start selling to a prospect.

Okay, cold calls are a necessary source of leads that should not be overlooked. Admittedly, some construction salesmen regard them as a necessary evil. The physical act of cold calling is often combined with another lead source. For example, the lead may come from a personal contact, and then the cold call is used to see the lead. Granted, this method should be used only as a last resort. If you have the lead's name, it will be possible to call and arrange an appointment. In my experience, when I have combined cold calls with any other source of leads, it has been because I didn't know enough about the lead to feel comfortable calling ahead; but this rarely happens.

Maybe I should give you my definition of a cold call before you skip the rest of this section. A construction cold call is one in which, no matter what the reason, it has been determined that someone may be in the market for new construction of some type, new or add-on.

Please be advised that I'm not suggesting you start at one end of the street and proceed to the other end, knocking on every door as you go. If you're doing anything at all, there's just not enough time; besides, that is a very unproductive use of your time.

I've worked out a very special use of the cold call, which you might try: I use cold calls as time-fillers during the day. I never set aside a certain time to make cold calls; as you well know, in the construction business too much happens on short notice that requires your attention.

For example, let's say I have a 3 P.M. appointment, and I've just left a job I'd been checking at 2 P.M. That gives me an hour, which may be lost time; so I'll put that hour to good use. I'll visit a part of town that I haven't seen for a while and just look around. I simply cruise about observing the businesses, looking for the telltale signs of growing pains. When I spot a likely looking lead and there's enough time, I go right in and start asking questions. Sometimes, because of time constraints, I have to return to make the call; but the lead is investigated.

I try to make approximately five cold calls a week; some weeks I make more, and some I make fewer or none. The construction salesman should always keep the cold call in the back of his mind and not only plan to make it, but do it! Cold calls require discipline. It is easy to put them off. I know; I do it sometimes. It takes only a weak excuse not to make the cold calls you know beyond a shadow of a doubt you should be making.

A construction cold call is a qualified call; you don't stumble around trusting to blind luck. The successful salesman knows like the back of his hand the signs that show when a business is growing. He's always looking about while driving, rather like a hunter after game. Common sense tells you what to look for, such as old buildings; materials stacked outdoors that really should be under cover (industrial firms display this telltale sign); jammed loading docks and truck parking areas; in the auto industry, many more cars surrounding a building than seems normal (this is a good indicator for auto garages and allied specialty shops such as radiator or body shops); trucks loading and unloading on the street through a front door with goods stacked on the sidewalk.

The details will differ for different types of businesses, and there may be special details to look for in certain areas, say in the rural business community as compared with the metropolitan segment. But one point will not change: there will be some sort of telltale sign of growing pains, and that's your tip-off to make the cold call. I'd like to add that a little practice will go a long way here; and the beauty of it is that you can do it while going from place to place, which gives you maximum use of your time. Time is the one thing you'll never have enough of when you really get into selling commercial projects.

You may wonder why you should walk in on Mr. X; why not jot down the name and address of the business and call? There is one good reason: the fast brush-off—and not because Mr. X is rude and hard to get along with. Oh, sure, there are plenty of those people around; and let me say here that the salesman who lets himself get pushed around by them won't last long. There is absolutely no reason for taking abuse. But, enough preaching.

The brush-off I'm speaking of is the kind every salesman gets if he's out there doing his job. Our Mr. X is busy; he's up to his knees in alligators when his secretary tells him there's a Mr. Somebody on the phone about a new building. Remember what we said before: the lead or prospect will always put his customers first and you second; so Mr. X gets on the line and tells you he doesn't have time to talk now, but maybe later. Keep calling and catching him busy, and suddenly you're a pest.

Well, you ask, what's the difference between this call and one you are following up from a personal contact? It's simple: most of the time with a personal contact call you have an entry, the mutual friend. When you get Mr. X on the phone and mention you're calling because Jack Smith suggested it, then Mr. X will be more inclined to give you a minute or two. You aren't a complete unknown; there is a line of contact between Mr. X and the strange voice on the phone. Take that line of contact out of the conversation, and you're next to nothing; the only way to overcome this problem is to let Mr. X have a look at you. At the same time, you can look Mr. X over and try to learn all you possibly can. If Mr. X is busy, you can wait until he has a minute; you'll be able to alter your game plan to suit the situation and make yourself more acceptable. Most businessmen will take the time to speak to someone who takes the trouble to call on them. So you're able to get to Mr. X and start qualifying him to see if he can be moved into the prospect column.

In the event that Mr. X doesn't have the time to see you then, at least you have a slight entry if you prefer to call before you try to talk to your man again. I know it's not much of an entry; but, believe me, it's better than nothing.

Of course, the lead can be checked out sometimes without talking to the person in charge. If for some reason you aren't able to see Mr. X, talk with his employees—the number two man or even the secretary. Just a simple statement, such as "Mr. X has just signed a new five-year lease," or "He's taken on a new line and is really going to need some more space," will let you know how to spend your time on this lead.

The best way for a construction salesman to save time is to find out immediately if the lead is going to build. Knowing when and with whom not to waste time is the surest manner in which to save time.

The cold call is a source of leads that has to be worked physically by the salesman. You should be aware of two facts about cold calls. One, they usually are long-lead-time projects simply because Mr. X probably has done very little to nothing about the project he's thinking about. If the owner has proceeded to get started, information begins to be known on the street; and the construction salesman may come by the lead from another source. So be prepared to use up a lot of time waiting for the project to move.

Second, cold calls are the poorest and least productive way to locate leads and then turn them into prospects. That's right, the worst way possible when compared with all the other sources. I'm centainly not telling you to forget about cold calls, nothing of the sort; they do have their place in the overall scheme of lead gathering.

The above statement is based on the hard facts of leads to prospects to contracts percentage results. The cold call is poor because you know nothing about it except what you can personally see; it's not like an incoming call from the Yellow Pages, where the simple fact that Mr. X has called you expresses interest on his part. The redevelopment and relocation project leads both have owners who have to move. You even have some degree of interest on the part of the lead with personal contacts and referrals. But with the cold call, you have nothing until you go in and find out. Simple mathematics tells us it takes a large number of cold call leads to generate prospects that turn into contracts. The cold call contract takes more work as a whole, but it's still an excellent source of leads and should be used.

Remember: learn to observe Mr. X's place of business, and then learn to notice changes that take place over a period of time. Always make it a point to keep returning to the business location which you feel offers cold call leads, so that you might see the changes that take place over a period of time. Also, remember, concerning cold calls, to get out there and diligently make them!

DISASTER LEADS

We've all heard the old adage "It's an ill wind that doesn't blow somebody some good." Well, there have never been truer words with respect to construction salesmen and disasters involving commercial buildings.

Fire is by far the main disaster that strikes buildings, explosions are next, and then Mother Nature can get upset and unleash a tornado or hurricane. No matter what destroys or damages a structure, get yourself in the picture as soon as you can, considering the owner's personal feelings at what must be a trying time for him. Every once in a long while you'll encounter the fellow who smiled all the while the firemen worked to put out the fire; the fire solved his problem. He takes the insurance money and builds a new facility where the old one stood.

Ninety-nine times out of a hundred it's imperative that the owner start a new building or repairs as soon as possible. Here's a good example, which I feel shows what the aggressive salesman can accomplish when he goes after a prime disaster prospect.

Back in 1971 I worked for a design-build construction company that employed another salesman as well as me. My colleague was working with a man in the furniture business who wanted to open a new branch. The lead, by the way, was a personal contact, which came from a club meeting.

Anyway, this salesman worked his behind off to sell the job. There was another club member after the job also, and the owner chose the other company's proposal over ours. I remember the reason—the other company's price was a couple of thousand dollars lower; and apples equaling apples, the owner opted to save the money. That's what the owner expected, anyway; but well into the job it suddenly came to light that there was quite a lot of fill dirt needed, and the cost wasn't included in the contract price. The end result was that the owner paid more than the price my firm quoted, since we had included the fill.

The owner was good enough later to admit he should have paid closer attention to details and signed with my colleague.

Six weeks later, on a windy Saturday afternoon, the furniture man's main store burned to the ground. That night my colleague called offering to help in any way, and this time he did build the new building.

This is also an excellent example of remaining objective and not getting personally upset when you lose a job. You never know when that owner will become a prospect again!

Disaster leads can be good ones because the prospect will have to have a building to operate from. Follow the news media, and pay close attention when you hear or see news that may give you a lead.

BONUS TO EMPLOYEES; FINDER'S FEES

Make salesmen out of all your employees. Stress to them the importance of bringing every scrap of information that may be a lead to the attention of the proper person. Instruct your people to keep their ears and eyes open. It's positively amazing what they can bring in when they are on the lookout for leads.

Now you can give the employees the old college pep talk about how much the leads mean to the firm and what's good for the firm has to be good for them. When the first lead comes in from your young bookkeeper and ends up in a big fat contract with a good margin of profit, you could give her congratulations and a big thank-you in front of the whole office staff. Months later you might still be trying to figure out why the hired hands haven't brought in any new leads since that last great one.

Well, nothing in this old world is free. Your employees are paid to do a job, nothing more. Why should they go out of their way to put money in the boss's pocket? Their check arrives regardless of the leads. The answer is to make it worth their while. Give the book-keeper a big thank-you and a crisp new hundred dollar bill in front of the office staff, and notice how there'll always be a lead or two coming from the help. Nothing else, absolutely nothing, motivates like money; so use it to make money.

Don't be selective; let everyone who works for you know he or she has the same standing offer of x amount of dollars for a lead that ends up a contract. Here's the catch: the lead has to become a contract. This idea of a bonus to employees will work and provide the firm with a new source of leads because seldom will the employees associate with the same people that the owner does.

Another paid source of leads can be finder's fees. In this case you have an understanding with salesmen in other and allied fields, say equipment salesmen, or insurance people, or maybe a commercial real estate broker whom you will pay a set amount for a lead that turns into a contract. These people are in and out of a great many businesses and can see and know what's going on. For example, why are certain business people talking with the equipment salesman about a new ten-ton bridge crane? Why are they talking with their insurance agent about the cost of their coverage if they happen to add fifty men to the shop payroll? Why are they talking to the real estate agent about buying the adjoining property? It would take a real dummy not to be able to see business expansion here.

The problem comes from the fact that others have knowledge that you lack. Realize, please, that you might get your lead as the result of a personal contact, and that's great. But personal contacts are personal; the person with the lead is doing you a favor by telling you—there is no incentive most of the time except for helping a nice guy or a friend. If you could only know the leads that started out to be passed on to you, but fell through some crack for one reason or another, you would just be sick—believe me. For every one you get, I personally feel three or four get dropped.

The answer to the situation could be a finder's fee. Nothing else motivates like money. But you can't go around offering money to your friends for leads. You can, however, offer it to people in business whom you see mainly during business hours.

I personally use the finder's fee offer very sparingly because I don't like the idea of being known for having to buy my leads.

This really is a personal judgment situation for the construction salesman. In other words, it's hard to tell someone how and when, if at all, to use the finder's fees. There may be accepted business practices in your area that can help you decide on their use.

My own guideline is that when someone brings me a deal, not a true lead, and it puts me in the know, then I'll build in the fee, I'm talking about one or two hundred dollars. For example, a person brings me a possible lead, such as that he knows about a new building, and he's trying to sell some part of the overall picture to Mr. X, and could he make a few bucks if he gets me in to see Mr. X. In other words, the lead-seller is shopping before the fact, and that's only good business—for me and for him. This is the kind of lead on which I like to use finder's fees. When I know what the situation is and have some measure of control, rather than broadcasting to the whole town I pay for leads.

In closing, bonuses to employees are an excellent source of leads, and I wouldn't hesitate to use them. Finder's fees I'll leave to your discretion.

ZONING CHANGES AND DEED TRANSFERS

Watch the paper and/or check with city hall for applications for changes of zoning to accommodate commercial enterprises. These are good leads because someone is planning to build and is going to considerable trouble to have the site rezoned.

Deed transfers are another good lead source. When Mr. X puts out hard cash for land, you know he's serious. Often deed transfers are not recorded in the local papers. My experience has been with four cities, all sharing boundary lines; two don't record deed transfers, and two do. I'm sure it's different from locality to locality, so you'll have to find a way to keep up to date on who buys what. Maybe your area will have a publication like the one available where I work. It's a booklet that the businessman can subscribe to, which lists deed transfers as well as law suits, who is filing for bankruptcy, and so on. It's a very handy item to have around.

These two lead sources can be pursued by your office help. Let someone be assigned the job every day to check the local papers and the publications, if you can subscribe to something like this, and mark the likely looking leads. The important thing is for the checker to cull out the useless ones, leaving the salesman the job of boiling down the original list. Every now and then a lead will come from a news item, and these should be brought up also. You'll go for days with dead ends and then bingo! up pops a good one; and this source of leads really doesn't take too much of your time.

RECORDS

The world's worst record-keepers are salesmen. Most hate paperwork, so they have a tendency to do as little as possible and operate out of their shirt pocket. I'm no different; I fit the description I've just given to a T.

It's quite important to have and keep up-to-date records to make sure that no lead or prospect is overlooked and allowed to get away. The very fact that long time periods are an integral part of selling construction projects makes good records mandatory so that the prospect can be contacted on a regular basis according to his particular situation. It's equally important for the owner of the construction company who has a salesman working for him to know just what is going on. Employees leave, and there must be some record so the job can be carried on.

I've found that a loose-leaf binder works the best for me. It's easy to add or take out pages as necessary so that the notebook is not cluttered with old, dead information. I keep a list of leads with just the barest information. My prospect list is just the opposite. I record every bit of information and the next step in the sales presentations to be followed. It may be calling Mr. X next month to check on a site plan, or it could be checking with the city after a meeting for Mr. X. The important point is to follow up on the information.

It does no good to keep records if you don't use them. I'm into mine every day making notes and making sure nothing is being overlooked. Failure to follow up tells the prospect you're not interested in his business, and when he gets this idea, you won't have it.

I might add that if there are any habits or strange things the prospect dislikes, write them down, and make sure the people in the office know they should check his sheet before contacting Mr. X for any reason. We're dealing with individuals, and everybody is different. Record those funny little peculiarities, and save yourself some embarrassment and maybe a job as well.

A few years ago I worked with a prospect who didn't like for a secretary to call him and ask for him to hold for so and so. He felt if so and so wanted to talk with him, he could jolly well ring him up. I made sure that when anyone called him, the person who wanted to talk did the calling. I discovered this by being in his office when he received a hold-on type of call, and he told the caller to have so and so call himself if he wanted to talk.

So write it all down and use it; you'll not regret the time and effort.

WHOM NOT TO WASTE TIME WITH

I said several pages back that the best way for the construction salesman to save time is not to waste any. You might call this the negative part of this chapter, since I would like to point out the type of leads you should not be pursuing. Now I'm speaking from cost, both in money and time, and experience; hopefully you'll be able to benefit from my mistakes—and I've certainly made some beauts.

Committees and Groups.   These people, with the best intentions in the world (since they will be nice people trying to do as good a job as possible), will bankrupt you if you let them. I'm talking about church, social, civic, and even theatrical groups. I firmly believe the first time in history that three or more people met together, the first topic on their list was a building to house their enterprise. It still has to be the main subject of committees to this very day. I'm always ducking them, and that's what I intend to keep doing—staying out of their way. It took me five years finally to realize that it's almost impossible to sell a group; and when such a monumental event does happen, there's bound to be some member who has a brother-in-law in the construction industry. This relative could be a plumber's helper or drive a truck for a supply house or even pound nails, but suddenly this unknown factor is brought up with near reverence in the meetings, and makes the brother-in-law that sits on the group the greatest authority on construction since the Romans. This expert, who doesn't know a 16-penny nail from a piece of angle iron, is the person questioning your every move, and worse, lots worse, questioning your invoices. His every sentence starts with "My brother-in-law says . . . , " and after the umpty-umpth time you start working on a way for a concrete truck to back over him.

No, my friends, it's not worth it; avoid groups that want to build a building. You'll live longer and feel better while doing it.

In my opinion, the real problem is that no one can say yes or no. Decisions take too long or never come at all. This situation applies not only to negotiated design-build projects but to bid work as well. I know a great many general contractors who hesitate to even bid group jobs because of this same nightmare. Then, again, there are some contractors who will do the job if at the outset it's understood the contractor will work through one person; and here the architect on the job can be a lot of help as the owner's representative. My opinion is that even these jobs become messy at times; the contractor just doesn't say much about it.

There may be an exception that you feel warrants your time and money. When this happens, keep both eyes open and be aware of everything you are letting yourself in for. Nothing else will grind up time and money like a committee that needs detailed information to start their ball rolling. If, and please note the *if,* you have to get involved with a club or church group for any reason, my suggestion is to do as little as possible, and try to get out of the whole deal as quickly as possible. Sometimes this is not possible; so limit the time and money you are throwing away.

I know all this may sound a little on the callous side; after all, these are nice people setting out to accomplish a goal. My answer is for you to please bear in mind your goal, the selling of construction contracts. *Selling* is the key word, and this means playing the percentages. The percentages for group projects are absolutely dismal. This fact is not something I made up; it's based on experience, and I still have the scars to prove it. If you work enough group leads, contracts can be had; the simple law of averages tells us that. But, the few contracts will never bring in enough money to pay you back what was spent working with all the leads that fizzled out.

Money is the bottom line, and this class of lead will cost you.

As I mentioned a couple of paragraphs ago, there are exceptions; so I'd like to tell you of my exception. That's right, believe it or not, I really have done *one* group job, and it was a good job in every way from start to finish. I doubt if I'll ever do another one because I know all my luck was used up.

In 1975 my lawyer called me about a lead. He asked for me to hear him out before I said anything. He proceeded to tell me that the local police department was ready to start construction on their local Fraternal Order of Police lodge, and he had recommended me to the boys in blue. I let him finish, then told him he had to be crazy. It's bad enough fighting with just ordinary people on a committee; I sure wasn't going to take on the local cops.

During the conversation it came out that my attorney was also the F.O.P. lawyer, and he was handling the entire project once the contract had been negotiated. Every time I came up with an objection, he shot it down. The reason he had recommended me to the police was that I had constructed his office building, handling the project from design to finish, and halfway through the job he leased out the extra space, so that I suddenly had to keep six lawyers happy while their offices were being finished.

Well, my attorney, who happens to be a good friend as well, held back the hooker for last. The police only had so much money, and it was imperative that they work with someone who was used to working within a budget. Also they wanted the building designed to be expandable to twice its size so that they would be able to hold dances. That set the hook. I enjoy a challenge, and this certainly was one; so I said okay and spent the entire next day wondering just how big a fool I'd been.

I worked with a committee of three to determine the requirements; then when the job started I worked with the chairman, who was a lieutenant on the force, and the attorney. The job went smooth as silk, and I admit I'm glad to have been able to help the police put the whole package together. It helped knowing that the paperwork was being taken care of in a competent manner, and the money was lined up and waiting.

This was my one and only exception, and I'd not even have considered taking the job if the circumstances hadn't been exactly as they were. There is one other good point—the police will be enlarging their building in the future; so there's a good prospect to keep tabs on.

This job has not changed my mind one bit about what I think about group or committee leads, which is: forget about them.

CONTRACT POTENTIAL

Up to now this chapter has covered where to go to find leads. But that lead is nothing more than a name until you talk with Mr. X and dig out the necessary information to determine just what it is you have to work with.

The good construction salesman would make just as good an investigative reporter, police detective, or intelligence agent. Why? Because people in all four of the above professions have to be able to obtain information vital to performing their job. Information is the job.

As a construction salesman calling on a lead, you are going to have to ask questions. Let me say right here, never hesitate to ask the particular question that you feel will help you. The very worst the lead can do is say no, and you're no worse off than you were before you asked; so you have nothing to lose.

I'm talking about legitimate questions pertaining to the project; don't be guilty of prying. As long as you're on the subject, ask away. An important fact to keep in mind when talking with a lead is that Mr. X may be able to pass on another lead whether or not he is a bona fide lead himself. Train yourself to bring the conversation around to your asking if Mr. X knows of anybody who may be in the market for a new building. You'll be surprised with the answers, and this type of lead from a lead will usually give you one that probably no one knows anything about yet, which puts you in the picture right up front.

Good information-gathering ability is not something you are born with. It has to be learned, and cultivated. When Mr. X tells you he doesn't own his building, you ask if the lease is short- or long-term. If the answer is short-term, ask how short, and explain that building a new facility is not a fast undertaking. It takes a lot of time; and since he has, let's say, eight months to go, why not let you work up some quick budgets so he (Mr. X) can start getting an idea of cost? If the answer is long-term, say 3 years down the road, ask if he has any plans for a branch operation. Don't just accept his answer about leasing and walk out. Dig about to learn something that may help you then or in the future.

In the next chapter I'll go into details of what to ask, but for now it's important to realize that the construction salesman must be able to ferret out bits and pieces from the lead. It's the only way to determine if the lead can become a prospect, and prospects are what you are after because they are the ones who buy buildings.

Let's quickly sum up this chapter, which is strictly about finding leads to investigate. The total number of leads that you'll be working with at any given time will be made up of leads supplied from the various sources I have discussed. No one area will provide you enough leads that you can afford to ignore the remainder. Temporarily, under certain conditions, you may get more than is normal from one particular class of lead, but don't let this lull you into a sense of false security. The picture can change, and you'll start out one day with no one to call on.

Keep all the sources constantly in mind, and pursue them. Remember, you'll be playing a percentage game and it's vitally important to remain 100% objective about the source of your leads. If there are things you personally don't care to do, then do them anyway. The reason you're doing this at all is for money; so don't be foolish enough to pass up leads that can provide the money.

Cold calls seem to be most construction salesmen's Waterloo. For that matter, cold calls in any field of selling seem to be the one least-liked part. I've even been told by some salesmen that they feel cold calls are somehow demeaning. To me, this is so much rubbish. What's the difference between a cold call and a clerk in a retail store coming up and asking, "May I help you," When the construction salesman makes a cold call he's asking the same question—maybe not in the same words, but nevertheless the meaning is the same.

The reason many salesmen don't like cold calls is that they are afraid of them—afraid of looking stupid, afraid of meeting a situation that they can't handle, afraid of being rejected, and so on ad infinitum, I imagine. No one likes something he fears. I recommend cold calling for a reason that is even more important than finding a lead. Cold calling will teach you to think on your feet, to handle situations, to meet complete strangers in all walks of life, and talk to them as well. In short, it will give the construction salesman confidence, without which he couldn't give a building away. And after you become comfortable with cold calling, don't stop even when you have plenty of work. It's a sure way to keep yourself fine-tuned to people. By the way, if you pick up a lead, then great; nothing like a good lead to perk up one's day, especially when you find that your closest competition knows nothing about it.

So wade right in, and worry more about making money than getting your hands dirty.

Lead gathering is plain hard work; please start out with that in mind. Ask questions and always be on the lookout for a lead from a lead. Never leave a Mr. X when you are through with your initial meeting without asking if he knows of anyone else who might need your services. It's truly surprising what people know but are not aware of until someone jogs them.

In closing this chapter I've listed a chart of the lead gathering areas along with their contract signing potential for the prospect. Believe me, I know it seems like a lot to remember; but after you have some experience under your belt, the chart will be imprinted in your head with your own variations added.

Lead Source	Contract Potential
1. Personal contacts	Good because lead has expressed interest.
2. Referrals	Very good because lead has expressed interest and taken the trouble to investigate.
3. Advertising	Good because lead has expressed interest.
4. Trends in specific business areas	Fair to poor because you know very little about lead at first. Will take many leads to produce good prospect.
5. Redevelopment project	Excellent because lead has to relocate.

Lead Source	Contract Potential
6. Relocation projects	Excellent because lead has to relocate.
7. Industrial commission	Fair because lead is just shopping around.
8. Rejected bids	Excellent because lead has everything together and is ready to move, but cannot for some reason—usually money. (Note: Move fast even at the cost of another lead if necessary.)
9. Cold calls	Poor because you know very little about lead at first. Will take many leads to produce good prospect
10. Disaster	Excellent because lead will probably need to rebuild to continue in business.
11. Bonus to employees; finder's fees	Fair because the person supplying the lead most of the time really doesn't know very much about the lead.
12. Zoning changes and deed transfers	Good because the lead is putting out time and money to do something.
13. Committees or groups	Zero; the kiss of death.

The above lead sources are listed in decreasing number of leads generated; that is, you'll have more personal contacts than referrals or more relocation projects than rejected bids. Now this rating system is very loose, but what I want to do is give you some kind of handle for judging your leads so that when you have to cull out and put priorities on them, you'll have some idea of just who's hot and who's lukewarm. In other words, don't break your neck over a lead from the industrial commission when you have one from a redevelopment project and only time for one.

Please remember that leads are the lifeblood of selling construction. They are where the prospect comes from; and he's the guy who hands you the check. So be on the lookout at all times for leads.

Chapter 4
Qualifying the Lead

The previous chapter covered the many methods that can be used to obtain leads. The problem with a lead is that it's nothing but a name on a list until you take the time to check it out and qualify it to see if the lead can become a bona fide prospect.

Not every lead will become a prospect. Again it's percentages; a lot of leads will boil down to a few prospects. I have absolutely no idea what sort of ratio you're supposed to get when comparing number of leads to number of prospects. There are way too many variables, and I feel a statistician would be hard put to arrive at a number. This much I do know: a lot of leads will give you a few good prospects. Please note the word "good." Now a lot of leads can give you a fair number of prospects; but we're not interested in just any prospect—we are after the good ones. These good ones are the leads that you prove to yourself are going to make a move; they plan to build. Granted there may be problems that must be overcome before the contract is signed; nevertheless, they plan to build the facility. That's my definition of a good prospect. All the others are nothing but people for you to very nicely contribute your money and time to. So I say again: A lot of leads will give you a few good prospects. So please be prepared to work for your prospect list. Oh, every now and then you'll back into a good one without much work, if any; but that's going to be the exception, not the general rule.

When qualifying a lead, use the direct approach. Come right out and ask Mr. X what his plans are; does he intend to build? Now here you will receive one of two answers, no or yes. If the answer is no, then you know right away this particular lead is a dead end; but never leave without asking Mr. X if he knows of any other likely people whom you might call on. Remember, never pass up a chance to fish for a lead.

The no answer is cut and dried; you know immediately where you stand. When the answer is yes, then you think you have a real honest-to-goodness prospect to work with. Well, I hate to burst the bubble, but this isn't necessarily so.

Our Mr. X who's just answered your question with an emphatic *yes* could be telling you yes while in truth he doesn't have the foggiest idea of what he's doing. He may have a burning desire to build that new facility; and just because he wants to, then he thinks he can do it. It's up to the construction salesman to dig out enough information from the lead to determine if he can really build, and therefore is a good prospect.

You won't know when the lead answers yes if the yes is good or bad. The best method to use is to assume it's a bad yes and start right off digging for the needed details. If the yes is a good one—that is, Mr. X has the preliminaries put together, and it's quite obvious to the salesman that this lead knows what he's about—then you lose nothing at all at the beginning by assuming he's a bad yes. You will quickly pass through the preliminary qualifying and get down to the nitty-gritty.

On the other hand, most of the leads will take qualifying starting at square one; so we'll start at the same place.

SITE

The A-Number-1 requirement for a contract is for the lead to have a site. Absolutely nothing will happen unless our Mr. X has a piece of land to put the building on. This is as important to selling construction as simple basic addition and subtraction are to the science of mathematics.

I stress this point so strongly because I've noticed that construction salesmen have a tendency to become so wrapped up with a project that this simple, basic rule is overlooked, whatever the reason or, better yet, excuse.

While you're talking with Mr. X, you steer the conversation around to the site, and Mr. X tells you he doesn't have one yet; matter of fact, he's not really started looking around, but he's going to do this and that when he locates the right spot and so on . . . . When you hear words to this effect, then get out as quickly and as politely as possible, and work on something else. Now I'm not saying drop this lead; I'm saying back off until the lead is serious, and lining up a site is the first step that shows that he is.

You keep the lead on your list and check back every few months to express interest and keep your name in front of Mr. X. After all, the lead can suddenly become serious and buy a site, and you'll want to know about it from Mr. X, not the grapevine well after the fact.

I'm sure you're aware that nothing is simple any more; so there are variations of this no-site rule to know about and to know how to use in your selling. Obviously you should react as in the previous paragraph if it's apparent that the lead has no site and is daydreaming; but what about the lead who has no site and who you feel is not daydreaming? He's going to do something. I want to be very clear on this point. I'm not saying forget a lead with no site; I'm saying don't spend any time or money except what you can easily provide with no trouble to yourself.

I have an excellent example of what I'm talking about. A lead came to me in the form of a personal contact. The lead was thinking about building a new facility; so I immediately jumped on it. It didn't take me long to determine that Mr. X was going to build; the problem was that he was undecided about just where to relocate—hence no site. I told Mr. X there was little I could do until he had a site because the prices would not be good indefinitely. He did give me the general specifications for his new store and asked for a handle on the cost, just a ballpark price. I have done quite a few jobs like the one he wanted, so it was only a fifteen-minute job to budget the cost. I didn't tell Mr. X this; I left him with the idea that it would take a little time but I was willing to do it if I could work it in during the next couple of weeks. I wanted Mr. X to start getting the idea I was really putting in time to help him. I was simply selling myself. Now that's all the pricing I was going to do—I didn't care what else was asked of me; nothing more would I do until the site materialized.

I returned with the budget a few days later, spent several minutes going over it, and then left, telling Mr. X when he found a site to let me know and I'd be glad to look at it for finding hidden construction costs. This would keep me in the picture and make Mr. X start to be dependent on me.

Six months went by; then Mr. X called me to check out a site with him. I might add that during the six-month period I had called him several times to touch base so that I could stay in the picture.

Well, we checked out the site, and it had nothing but problems—sewer, setback requirements, plus a road exposure that was terrible for a retail business. Site costs would run up the overall costs, and maybe he should look around again, I told him. This went on for a year, and all I had invested was a little time. I felt under the circumstances that the small amount of time was worth it. At no time did I do any more estimating. I was positive Mr. X would build when he found the correct site. Also he was calling me, which showed interest. Well, after a year he finally located a site that filled the bill, from his business viewpoint and with respect to the construction cost. The real estate agent was a friend; so I asked him to keep me informed as to progress made on buying the lot.

Three months later the agent called and told me Mr. X had just closed on the site. You know, this brings up a point that's a complete mystery to me. Why does it take so much time to settle something that's on the market to be sold? Land deals seem to involve no concept of time. They're as slow as bank loan committees. Anyway, with the site in hand, Mr. X became a prime prospect, although granted I had somewhat of the inside track by virtue of my prior work. To make a long story short, it turned out to be a nice job.

My point is that Mr. X didn't have his site but did meet all the other requirements; so I continued to follow up and work with him as long as the costs to me were rock-bottom minimum. I wasn't going to provide plans, price letter to banks, and all the other help necessary to getting a project off the ground.

I mentioned earlier bringing the interview conversation around to the site. Just how the salesman goes about asking for information is important. Think back to the first chapter and low profile/soft sell. Now with this in mind, imagine how you would look to Mr. X if, when you asked him if he had a site and were told no, you then closed your briefcase and turned to go, saying over your shoulder, "Call me when you have one."

Qualify a lead like that, and you can bet he'll cross you off his list.

The way it should be done is to build the correct question right into the flow of conversation. For example: "Mr. X, in order to work up a meaningful price, I'll need some information on your site." Then you give the reason for needing the facts. Please remember Mr. X will not be familiar with all the little details but will start to realize there's more to building a new facility than just saying, "Let's do it." You are selling Mr. X on your competency. So you go on, "It's important for me to know where the sewer is located so I can check with the city to see if there will be any problems tapping in."

The reasons given to Mr. X for asking for site information can be numerous; just use something. Surprisingly enough, I've found that most of the time there really are legitimate questions which need to be answered.

At this point Mr. X tells you he has not located a building lot. Don't let on for one second that warning bells are ringing inside your head; calmly go on with the meeting. Move ahead to the other qualifying questions. Make no hasty decision yet concerning this prospect. You need all the facts first; so go ahead with your diggings.

After you're sure nothing will happen until Mr. X locates a site, then you have to pull back without giving the impression to Mr. X that that's what you're doing. The tack I always take in this situation is to tell the lead that until a location is picked out, it's very difficult to work up any costs that will be worthwhile. I tell him construction prices are always in a state of change. It will do no good to supply building costs without site costs because the bank is interested only in the total cost. Then I throw in the hooker: I tell him I would be doing him a disservice to give out what has to become worthless and, worse yet, dangerous information.

At this point I offer to give him a handle on the builing cost if he will use it only for personal information and take note of the fact that it will be on the high side. Now I offer this mainly to keep myself in the running if, in my judgment, this lead has promise. This judgment call is just what the word says; it's strictly based on my feeling about Mr. X, how he comes across, what has been done to date—in short, the sum total of all the small parts that come out in a qualifying interview. Here the construction salesman is 100% on his own. He makes a decision and then proceeds from there, often hoping he's done the right thing. This is quite hard to get out of a book. I can try to give you some guidelines, but that's all; the rest is up to you. And if you're working at it, you'll make mistakes. Don't let them worry you. Mistakes come from doing something; mistakes aren't made if you're not doing anything.

Sometimes "no site" can be used to help sell yourself to Mr. X. I think many times our lead hasn't done anything about his site because he really doesn't know how to go about it. Oh, he has some ideas, but he may hesitate to become involved with a real estate broker. Many times he's so wrapped up in his business that he won't take the time to drive around and see what's available. Here the construction salesman can be of some help. Tell Mr. X some of the details concerning water, sewer, gas for heating, and so on, and then recommend a good commercial broker to work with him. If he says okay, he'd like to talk with him, offer to call the broker yourself; tell Mr. X that way you'll be able to give the broker construction facts which will help the real estate man do his homework so as not to waste Mr. X's time with unsuitable locations.

This strategy does several things for the salesman; it lets him stay in the picture and also enables him to call the broker with whom he wants to work. You give him all the information concerning Mr. X, if possible even going so far as to give the broker your readout on how Mr. X likes to be approached. The really important fact is that you have called and given the broker a good lead to work on. Since all of us travel a two-way business street, the broker will want to repay the favor in the future.

One more excellent point to bear in mind about contacting the broker will be the possibility of developing an ongoing relationship that will keep you informed about the progress being made on finding the site. He might even learn if another contractor comes into the picture and pass that information on to you. Forewarned is forearmed. Be sure to ask the broker to contact Mr. X right away and let him know that he (the broker) is starting to work on his site. Then several days later check with Mr. X to find out if the broker has indeed made contact with him. There are two reasons for checking back with the broker: one, to keep the ball rolling, and second, to show Mr. X that you're a hustler, which is just another small step in selling yourself. Never let up on selling yourself; it has to become as natural as breathing.

Okay, let's quickly sum up. Without the site there can be no contract. If the lead doesn't have his site, then be careful about spending time and money on him. Do offer to help by bringing in a broker (whom you can work with); then step back out of the way, not out of the picture; and keep tabs on the situation because when Mr. X obtains his site, then it's time to go to work.

FINANCES

The second most important fact in qualifying a lead is money. I've always heard that love makes the world go around. I don't buy that—it's money!

Without money our lead is a dead end. It will take money as well as a site to make this lead into a prospect who can sign a contract. There is one major difference between the two. The site location can be critical, not any old place will do, while in reality any old money will do nicely.

Bringing up the subject of money in the conversation is a little tougher than bringing up the site. The best approach to opening the door on the money question is to assume that Mr. X has no problem at all. He will need some information to satisfy the lending people; so you offer to help with what's needed. This will usually start the talk about finances, and from there it's an easy matter for the salesman to ask some direct questions under the guise of gathering information to help. You are doing just that, but at the same time your prime concern is whether Mr. X has his ducks in a row so he can pay the bills.

Nine times out of ten the lead will have talked with his banker and will have some idea of what he can expect. Usually a good businessman will talk to his banker first, before proceeding any further. Now without plans and contract costs to supply to the lending people, everything is verbal and tentative. Nevertheless, the lead will have some idea about the money. My experience has been that Mr. X won't hesitate to tell the construction salesman about already talking to his banker and that there is no problem. Now I've also found that's just about all Mr. X will say about his money; he has it lined up, and that's all he'll tell you until he knows more about you personally.

The gray area is what you have to be prepared for now. It's sort of like the twilight zone where it's very hard to prove something that's important to you. Mr. X tells you he wants a building and you ask for details, which are forthcoming. Mr. X tells you he has has a site, and you ask where; the answer is a simple address—you are dealing with something tangible. But, how do you prove to yourself that Mr. X has the money arranged as he says? Start prying too deeply into his finances, and you'll probably push his off button (remember that from Chapter 1), as well as make Mr. X somewhat upset. No one likes to be doubted with respect to money; so you'll just have to accept his word.

The only defense you have is to try to find some circumstantial evidence that could give you some idea of whether Mr. X indeed has the money. The best way is to offer to supply some needed information that the lenders will require. Sometimes this will start our Mr. X discussing some of the financial aspects. Hopefully you'll pick up enough to put your mind at ease. In any case, you'll be in a gray area until the project moves to the plans step. The good construction salesman learns to live with this problem; it comes with the territory. It certainly won't hurt if you are able to make some discrete inquiries as to how Mr. X stands financially within the local community.

Mr. X has the money, or at least you have been told as much; but what about the lead who has a site and no money? You'll come across this situation every so often. I'm talking about a lead who wants and needs a new building, has optioned a site, but for some reason can't get up the money to make the deal fly. The lead will probably not hesitate to tell you his tale of woe, for he's desperately casting about for help with his problem. This type of lead can be a good prime prospect under the right conditions.

The construction salesman will have to know something about design-build and then leasing the new facility back to Mr. X. I'll cover the details of how this is done in the next chapter. For now, we are only interested in qualifying the lead.

You may be asking, why not assume that the lead has the money to build and not use this as a qualifier? Well you can, and a great many times it will be perfectly all right. But you have no defense against the big talker who likes to think of himself as a high roller. I use a handy rule that is accurate almost every time: Beware of the easy sale and the lead who casually brushes aside talking about project finances.

The money is why you're out there talking with this lead; so find out all you can. But at the same time remember this is probably the touchiest subject you'll discuss with Mr. X.

REASON FOR BUILDING

Our man now has a site, and his money; so let's look at another important area. Why does he want to build? Think back to Chapter 1 and why businessmen build new facilities. They are forced to. They may be forced by being in the path of redevelopment and relocation projects, or they may be forced to expand because their competition is expanding and they are afraid of losing business. It could be that their lease is up, and they can't renew it. They may have no choice but to build the new warehouse to be able to handle the new accounts they've just taken on. Regardless of the amount and type of force being applied to the lead, Mr. X will only move when he has to.

It's important to the construction salesman to find out what the moving force is. It will help him qualify the lead, to see if there is a good prospect to be had, and also help the salesman determine the time factor involved if the lead does indeed look like a good prospect.

The following example shows exactly what I'm talking about. In August 1977 I contracted with a local firm to build them a new office building. I had picked up a very general lead from a friend in the summer of 1976. The general lead was that the local Air Force base was expanding the safety zone at the end of its main runway. Thus, I had learned of a relocation situation through a personal contact, so that two lead sources were actually combined. I felt a little sheepish about finding out about the relocation project this way; I should have been on the ball checking with the base more often. Anyway, there was only one good lead to come from this area; so I went right to work. The lead checked out okay in every qualifying category, with one small exception concerning the reason for building.

There would be a time lapse until the Air Force actually paid for Mr. X's building, and he was not going to do one thing until he had the money in hand. With this information I told Mr. X there was no need for me to do anything until the Air Force closed the deal. We left everything as it was at that time with the clear understanding I would stay in contact to follow up. It took almost a year for all the final details to be worked out, at which time I went to work to put the job together.

As you can see, it's vitally important to know the details of why Mr. X has to build. It helps the salesman judge exactly how to proceed with the project, and at the same time the salesman can determine just how serious a lead is. As in any other selling business you get the lookers, the people who may be thinking it would be nice to have a new building and who want to know what it would cost on today's construction market—which leads us into the next topic.

INTEREST

There is a way to tell if a lead is really serious or out window-shopping. There's not too much to it: you give the lead some small duty to perform. He's serious if he does it, and not serious or at least questionable if he doesn't do what is asked.

The salesman should be able to get a feel for the lead as far as interest is concerned; this will come with some experience. It helps if you can have a readout before you have to "assign" a duty to the lead. The problem is that this requires a return call or maybe several until you can make your determination.

Okay, you say, what do I do? Simple. Ask the lead to get you some information on the site from the real estate broker, or the surveyor if one is involved. Ask him to check on a zoning question, or perhaps a decision about some detail of the building that you need to know before the price can be estimated. Here's a pet method that I've found most reliable: I ask Mr. X to give me an hour's time so I might show him some buildings for him to get a better idea of what we're talking about. If this is really impossible, and it will be with a retail businessman once in a while, give him directions and suggest that he drive by on a Sunday afternoon and have a look. If he never does, then forget him.

Let me give you a good example. In 1976 I was trying to qualify a lead who thought he wanted to build a new retail store. Everything checked out A-okay: he had a site; the money was there. But Mr. X didn't seem to want this new store quite badly enough; something didn't ring true—so I "assigned" him a duty. I had been involved with two stores in his type of business earlier, one in 1973 and one in 1975. I explained in detail how these two buildings were different from each other, and still used for the same purpose, and that I wanted him to see them. The one I built in 1973 seemed to be exactly what he had in mind. I gave him directions, and it was only a forty-five-minute drive from his home; so he planned to take a look on the following Sunday. The next Tuesday I stopped off to see what he thought about the store. Well, he hadn't gone over to look at it. Seems he decided to stay home that Sunday.

Now I ask you, if you were going to build something, and you found out where you could see what you had in mind, would you go look at it if you were really interested? You bet you would! Mr. X wasn't interested enough to go look; so he must not have been really interested at all. I dropped the lead like a hot potato, and a year has passed. He's still not built, but there's plenty of talk still being generated. Who knows? He might build yet, but the odds just are not good enough for me to donate any more of my time.

The interest area is also a gray zone. It's quite hard sometimes to work up a readout on the lead; and, believe me, nothing is guaranteed in construction selling. I've found from experience that you will seldom disqualify a lead from lack of interest. I think every businessman would like to talk about a new store or expanding the warehouse. But it's important not to take interest for granted; when you do, your chances of being used increase. This is definitely not good for the construction salesman.

I follow a general rule on this subject and think about it enough to keep my perspective: Do all you can to maximize the number of good prospects, and do all you can to minimize the manipulators. You will be manipulated! It really will happen, and it will probably upset you, but don't let this unpleasant side of construction selling turn you off. It comes with the territory because you're dealing with individuals before they are 100% ready to build. The whole idea of qualifying the lead is to cut down on the instances of just being used. Remember, there is nothing at all which says a good prime prospect will not nail you to the wall.

When I first started selling construction, I was Mr. Eager Beaver and gave the lead/prospect everything he asked for, most of the time with no questions asked. My "being used" percentage was pretty high. As I gained more experience, this percentage steadily dropped.

Another factor has had a small part in this percentage drop: age. Now I honestly can't say much for becoming older except that it seems to help your business image. I will have more to say about both age and business image later. For now we are mainly interested in qualifying the lead.

There are two ways of being used to be on guard against. The first one is almost impossible to defend your self from. The lead checks out, and you start selling; you take all the information, do your homework, and put together a price. The prospect takes one look at the numbers, has a fit, falls in it, and discards the entire project. You know there's no way to cut costs and meet the requirements, and the prospect won't compromise; so there you are. All that work for nothing, but—and this is an important but—neither one of you knew this in advance; so I suggest you make it a point in cases like this to part friends because your defunct prospect may decide to make some changes later. Better yet, the construction salesman should keep in touch. A phone call every so often will do the job.

In the other type of manipulation the prospect is using you to obtain every scrap of information, drawings, and whatever else he can get. He knows all the while that he doesn't plan to build then for some reason or other, but he doesn't hesitate to take advantage of you. Sometimes a person will do this and see absolutely nothing wrong with it. After all, you did call on him and ask for the business; and since you offered, he intends to get all he can. This type of person, I've found out, will let you run your legs off but will do nothing himself. Assigning him a duty to perform will flush him out every time. It will not get done, and the excuses you'll hear can be somewhat entertaining.

Sometimes the interest factor can be combined with other areas of qualification. For example, the interest emphasis may be put on the site so that you're checking two qualifiers at the same time, which will give the salesman a good handle on the lead. There are other ways to combine qualifiers, all with the interest factor. Again, for example, if I get the feeling the lead is a blowhard, and he tells me money is no problem, I respond immediately with how glad I am to hear that; and say that since he's a good businessman, I know he will want any general contractor bonded who does work for him. That being the case, I'd appreciate a copy of his loan commitment for my bonding company so that there will be no doubt about the money back at the home office of the bond company. If the lead is really using you, he probably hasn't even talked to his banker, much less obtained a loan takeout letter. Mr. X at this point starts backing away. His hand has been called with some heavy stuff. He won't like hearing about home office bonding people being interested in him. He will realize he's dealing with a professional, not some rinky dink he thinks he can mentally manhandle.

I've used this interest "dig" quite a few times, and it works every time. I've been surprised also: twice the lead handed over a copy of his takeout. Don't get me wrong—I was delighted, and one of the two turned out to be a quite profitable contract.

I'm sure with time the new construction salesman will be able to feel comfortable in this area and have confidence in his decisions. You can't be right all the time; but when the percentages are totaled, you'll be ahead of the game.

Now remember, you're talking about business with Mr. X; so don't be afraid to ask questions about his business and the planned facility. You can never have too much information.

I hope this past few pages haven't "put you off." I know I've painted a grim picture, and it seems you'll be out there on the streets looking over your shoulder all the time. Not so! Everything has a price, and this is the price you pay in selling construction; but look at the other side of the coin. There will be many profitable projects done with first-rate individuals; some will even end up as friends. Let me give you a nice example. In 1972 and 1973 I sold three 18,000-square-foot roller-skating rinks to one owner. This owner is now a good friend, and every year my daughter receives a free pass good for one year at any of his rinks. (Of course, my friend is also a good businessman, since every time my daughter wants to go skating, she manages to fill the car with skaters who have to pay!) So don't be downhearted; there are plenty of good times in selling construction.

To sum up the interest factor: If Mr. X doesn't display any, then move on to the next lead; for without interest, there's no sale.

TALK TO THE PERSON WHO MAKES THE DECISIONS AND SIGNS THE CHECKS

This next qualifying topic is the one where, in my opinion, construction salesmen fail to use their heads. More time has been thrown away talking to and working on the wrong person than in all the rest of the qualifier problems put together. The name of the game is selling percentages; but here the construction salesman has a mental block, especially the new one.

I can understand perfectly how you can get caught up and carried along with the momentum of putting a project together. You make a call. Mr. X is easy to talk with, and he's planning on a new warehouse. You ask about the site, and he takes you out back and shows you the building location. You ask about the money, at the same time knowing the firm enjoys an excellent credit rating. You establish the reason for building—taking on two new lines of equipment. There's interest shown. When all these subjects click, the good construction salesman is like a racehorse in the starting gate. And when the prospect asks for a proposal as soon as possible because he needs the new building by such and such a date, then look out—our salesman is cranking up! It's exciting to put a beautiful deal together, and very easy to become intoxicated with it. So our eager salesman puts together a proposal, and presents it to Mr. X. Mr. X looks the information over and casually says, "Looks okay. The price is within our budget also." Our salesman is ready to whip out his pen so Mr. X can sign. Then Mr. X drops a bomb on him, "I'll get this right off to Mr. So and So; he's the man in our main office who'll make the decision."

At this point, our friend, the salesman, has just said goodby to a whale of a lot of work, time, and money. Believe me, this can happen to you if you're not careful. I know from painful past experience.

There will be exceptions. For example, the branch manager is a friend, and he asks you for a proposal. You don't mind working one up because you feel you'll have a friend to help you. Matter of fact, I have a proposal that was sent to a main office just last week, and it was prepared because the manager is a personal friend.

These exceptions will be few and far between. As for the rest of the leads, keep both eyes open, and know just what it is you're getting into. Let me warn you right here to beware of branch managers; they are the best manipulators the construction salesman can run across. I mean nothing personal; I have some close friends who are branch managers of out-of-town firms. It's in their nature to use the construction salesman because he can help them. They all want more space, office, storage, whatever, and they persistently hound the home office. When the salesman walks in and offers to give him some more ammunition to send to the home office, the manager takes him up on the spot. And I would do the same thing in his place, I'm sure.

The best defense against this is to find out right away just who it is you are talking to. Now use a little finesse here; don't push the off button. I've found that if I ask someone who obviously is not in charge who is, and then he turns out to be the owner or manager, he helps me through what could become an awkward situation. Anyway, be certain to whom and to what you're offering your services.

When you determine that Mr. X is indeed the manager, go right on with the qualifying and requirements if the lead is a prospect. Tell Mr. X you'll be glad to work up a proposal; then ask him when the man from the home office will be in town—it would help to know this so that you could schedule your time. Also be quick to add that if time is very important, you'll be glad to visit the home office. Now what you're telling Mr. X in a nice way is you're not going to do business with him. You expect to deal with THE man.

Let me say right here that I have never sold a job where I couldn't talk with the decision maker. I have sold several very clean, fast-to-build, and profitable jobs to out-of-town companies where I could make my sales presentation to the person who said yes and no. I'm speaking from experience; I was taken to the cleaners so many times that even my wife told me to wise up. Now

I don't make proposals to managers unless it's immediately clear that I talk to the decision maker or in exceptional cases to the manager, if he is a good personal friend.

You may feel that you're walking away from some business, but you're really not; for if you do luck out and get such a job, there's never enough money in it even to begin to pay you back for all the work put into the ones you missed. Percentages again—that's all selling is; and the successful salesman knows how to play them to his advantage.

Now I know you'll uncover a lead who turns out to be a super good prospect, with the only problem being you're dealing with a manager. It looks good, and you feel you should give it a shot. Go ahead, that's fine; except don't let it take away anything from a better prospect, even if the prospect doesn't offer the same inducements as the manager job. Look on it as investing some money in a high-risk operation. Invest only what you can afford to lose. Play the percentages; better yet, believe in them. A good friend of mine, who's an Air Force pilot, told me the hardest thing he had to learn about flying and remember to do was to look at his instruments and *believe them*, no matter what he thought personally. The same goes for the construction salesman; learn to read the signals and then, for Pete's sake, believe them.

There's yet another pitfall to watch out for: the father and son team. It's hard to determine whom to talk to. If you start out with the father, chances are you'll be okay; but if the son is the one wanting all the information, be careful. He has to be considered in the owner category, but chances are he won't be the one who will make the decisions. He'll also be eager to do something—he'll be interested in making the business grow; so he'll jump right on the services you can provide. Then the old man will say no for some reason or other. Be prepared also to say everything twice with the father-and-son lead; for some strange reason you can never see them both together—seems one always has to watch the store. Just try to find out all you can about the father/son lead. It's important to talk to the decision maker without making the other one, (usually the son) upset because he's not the one running the show. You've got to be an excellent diplomat in this situation.

To sum up, if you cannot talk to the person who will say yes or no and sign the check, you're wasting your time. This qualifier is just as important as any of the rest I've explained, maybe even more so; for it is the one that's the easiest to overlook and/or want to overlook, so that the salesman gets sucked in.

PERSONAL PROBLEMS THE LEAD MIGHT HAVE

This subject is not a qualifier in the same sense as the preceding ones, but knowing about the lead's personal problems will help the construction salesman make better use of his time.

Please note that I'm not talking about personal problems in the sense that immediately comes to mind; I'm talking about the personal problem that is talked about in public.

Here's a good example. I'm working right now with a good prospect who is planning to build a 20,000-square-foot metal building, 34 feet high, to be used for boat storage. In the course of working out the requirements, Mr. X mentioned that he wouldn't need the information for three or four weeks. Seemed Mr. X was due to have an operation and would be out for almost a month. With that information I'm able to work my schedule around the proposal to my advantage.

Most of the personal problems that I've had to contend with have involved sickness, and I'm sure your experience will be the same.

Sometimes the construction salesman picks up a piece of information that really can be a semi-qualifier. Always ask the lead what his time schedule is; usually here is where the problems will come to light. At the same time, if you hear something about Mr. X having trouble trying to buy out his partner, and that he'll not be free to pursue the project until that's taken care of, which should be in such and such a time frame, start growing a little cautious. Don't jump up and run, but become wary.

Remember to be tuned in to anything that will help you. Pay attention when Mr. X says something about personal problems. You may even be subjected to hearing about his upcoming divorce or whatever. If so, listen; you can never have enough information.

To sum up, a lead is nothing more than a name until you qualify that lead into a good prospect. A good prospect is one who passes all the qualifiers and plans to build.

The qualifiers that you will use are: site, finances, reason for building, interest, talking to the person in charge, and personal problems.

It's imperative that the construction salesman understand how these qualifiers are used and why. Their use should become second nature, and with experience it will be possible to qualify a lead in a general way without calling on him personally.

When a friend passes on a lead to you, ask if Mr. X has his site and money. Your contact may have the answer and save you some time. But don't depend on him too much. It pays to contact the lead even though you know he doesn't have a site, to show him you're interested so that you can keep your name up front. Also you may be able to recommend a solution to the problem. Matter of fact, a solution for the problem would be an excellent reason for contacting the lead.

Asking questions from every source, digging out details, and putting all the information together is going to be second nature to the really good construction salesman. And I'm sure you are just like me: you'll have to learn and develop these techniques. No one is born with them. Granted some people seem to be more people-oriented than others, and it helps; but it's not mandatory. Hard work is!

Learn how to qualify that lead; if you don't, all you'll be doing is spinning your wheels. Nothing else is as disheartening as working your behind off with nothing to show for it. This chapter will help you hold such wasted effort down to a minimum; it will also help you improve your contract-signing percentage by allowing you to work prospects who plan to build. All you have to do is convince them to let you do the job.

Read on. The next chapter tells you how to convince them.

Chapter 5

Selling the Prospect

The purpose of everything the construction salesman has done so far is to get him to this point. It's as if he has been in basic training, and the time has finally arrived for him to move up to the front lines.

From here on the order takers are in trouble, because it takes a results-getting salesman to convince the prospect to give him the contract for his new facility. The sorriest of salesmen can do with ease what has been gone over so far. It's really fairly simple to dig up a lead and qualify it. It mostly just takes hard work. But now it takes brains as well as hard work to sell the prospect. Prospects are what you'll be working with now, no more leads.

An old sales manager I worked with when I first started out, more years ago than I care to count, told me right after I went to work, "Never forget, boy, nothing, absolutely nothing, will happen until you sell something." I've never forgotten what the old gentleman told me; but more important, he's dead right. Nothing does happen until somebody sells something!

So now it's up to you to earn your keep. It's time to sell a job! And here's how you go about it.

SECURITY

Up to now, with minor exceptions, the construction salesman has not had to rely on any outside help. He's been a one-man show except for other employees with his firm. Now when he starts selling Mr. X, he will have to bring in outside help. For example, there may be a zoning problem the salesman needs to check; so he visits city hall and talks to people about the situation. He may see a health department official if the prospect is going to build a restaurant. Subs will have to be brought in to look at preliminary plans and quote prices. At times a sub will be asked to quote on a design-build plan that requires details about the prospect. In short, you are putting the job prospect on the street. Don't underestimate your competition; if they get wind of the prospect, they will be all over him. This will make your job of selling that much more difficult, as well as cut down on you chances of signing up the job. Percentages again: the more contractors working on a prospect, the worse your chances of selling the job.

I still recall seeing posters during World War II depicting a ship poised and ready to take the final plunge, with the caption "Loose lips sink ships." Well, let me tell you, loose lips can cost you a job also.

The construction salesman should concentrate on job security and not worry about what anyone thinks about being security-conscious. There are a number of ways to work the job without giving out any information about who the prospect is.

Preliminary drawings are usually a dead giveaway. It's only natural to cover the sheet with the prospect's name, if for nothing else than to impress Mr. X. Don't do it; fight the impulse, and leave it off. Also, while we are on drawings (and I'm talking about preliminary drawings or sketches that you will be using to get budgets from subs, not architect-prepared plans; they come after the prospect is signed up, and then security is not as important), don't spell out what the building will be used for. For instance, if you are working on a retail auto parts store, don't spell it out. List the job only as "retail store." That's hard to pin down; "auto parts store" can be run down by a smart construction salesman in a matter of hours.

You may ask, how does it leak from the sub? You have a good relation with the sub, and he's not going to run his mouth. I agree he may not, but chances are he'll have to contact a supplier; there will be salesmen calling on him so he can put together his price to you. Your sub may be able to keep his mouth shut, but he may run a sloppy office as far as security is concerned. All an outsider has to do is look around on the desk or drawing table, spot your drawing, and right then he's in the know, and your prospect is on the street. Contractors' offices, both general and sub, generate a lot of activity. There are people coming and going all the time, and anything left out becomes general knowledge very rapidly.

Security was a one-man problem when you were working alone; when you have to bring in an outsider, security is still up to you.

So, as we've said, don't let your drawings give you away. They are one of the best ways of revealing information, since it is recorded on paper to be read by all. A few simple precautions will go a long way in this case; use your head and don't be careless.

Also, when talking to anyone, including subs, refer to the job in general terms. The prospect simply becomes "the owner" when I'm discussing something that pertains to the job.

There is one particular area where I feel general contractors are extremely lax. The area is internal security. Please don't laugh; internal security is very important when you're doing design-build selling. The best rule to follow when you're trying to sign up a prospect is right out of the spy handbook—what they don't know they can't tell. Now I'm as sure as I can be that you don't have to worry about industrial espionage. What concerns me is the occasion when the steel foreman is having a beer with his opposite number who works for your competitor. Shop talk will almost invariably occur, and all your man has to do is say that his company is working like crazy to sell a new warehouse to a moving and transfer company. It would be said in all innocence, but the point is there's an excellent chance that the word would get back to your competitor. I know this sort of thing goes on. The people in a certain trade all know each other. They seem to float around working first for this company and then that one. Often these people will work together for a while, then for competitive outfits, depending on the work load. But they still know and see each other as well as discuss their work.

You want to make sure your prospect is not the main topic of discussion at the local tavern next week. Any employee can blow the whistle on you, and never be aware of what was done. The secretary going to the post office or the bookkeeper making a deposit at the bank can pass the time with someone from another construction company, and zap, that's it. I feel the real problem here is that construction company employees are a little more concerned about job security than people in other jobs; therefore the uppermost topic is, how's your work load? They're not prying, just interested. If one firm's contract load is off, they want to be reassured by hearing someone else's is also off, which probably means the industry is off a little. Misery really does like company. On the other hand if they are loaded with work, they want to brag some. The net result is that the work load is talked about often; and in such conversations the cat can be let out of the bag.

Now it's impossible to keep your employees in the dark. You can impress on them that their very job may depend on their keeping their mouths shut concerning prospects and upcoming jobs. Maybe it wouldn't hurt to suggest they listen a little closer to other people and bring in a lead so that they might be able to pick up a nice bonus if a sale is made.

I feel the best way to handle the internal situation is to explain to employees that what they hear in the office is confidential. Next, work on a need-to-know basis. If you have an estimator to work up the prices, make sure he doesn't run through the office telling everyone. At this stage of the selling procedure, he's probably the only one who needs to know anything specific. All the rest of the office needs to know is that the estimator is working on a building price. As the selling process moves ahead, other people will be brought in. The bookkeeper will have to charge off the estimator's time to something; so this person gets in the picture: Finally the proposal has to be typed, and suddenly the typist knows almost as much about the job as you do yourself; hopefully Mr. X signs the contract soon after this.

Along with the office help, you'll have the chief field people stopping by the office, and it's only natural for them to have a cup of coffee and chat—with you, or the estimator, and certainly with the women employees. You may stop by a job and find that the conversation turns to work in the pipeline because the foreman is vitally interested in the future workload. This will often happen if you're working as a construction salesman for someone else. Anyway, it's hard not to let the key field people in on what you're doing. Tell them anything you like as long as you don't compromise the prospect. Now this takes some diplomacy because you don't want the foreman to get the idea you're hesitant about talking to him. This is an area where I can't tell you what you should do. Only you know the foreman in question well enough to handle the situation.

There's a natural tendency to want to give your people a lift by telling them what is going on when the future looks bright, but please don't tell all if it might come back to haunt you. Learn to think about your external and internal security. Good prospects are not that easily come by; so give yourself every break to sell the job. The same sales manager I spoke about a few pages earlier told me something else that I'd like to pass along. He said, "You never learn a thing talking." And I'd like to add my own rule, "You'll never give anything away when you're listening."

For your own sake, prospect security is extremely important; so practice it.

PROSPECT LIST

Before you start out the door to sell a contract, it will help you a great deal to know how you're planning to use your time with prospects.

I've found the most efficient method is to keep a simple list of the prospects in order of their importance. A glance at it always helps me keep my perspective about whom I really should be working.

Your list will always be in a state of change, depending on the progress and, in some cases, lack of progress of the negotiations. Remember, your list should be made up of people who are planning to build, so that everyone who is on the list is a possible future job; and you should divide the list into categories, depending on just how good a prospect each Mr. X is. For example, I use three categories:

Fair—This prospect may have the desire; but money problems, zoning difficulties, or some other roadblock could be in the way. Nevertheless, the prospect may resolve the problem, or you may help in some way to move the project along. Because of the situation, though, this prospect is only a fair one. Reason: time consuming, long-range problems.

Good—This prospect has the desire, no problem with money, and even a site (in some cases there will be more than one site to contend with); but for some reason he is not committed 100% to moving ahead. The state of the economy is usually the reason, or there could be personal reasons.

Excellent—This person has his act together: site, money, and his mind made up to build! He may be the prospect who is forced to move. No matter the reason, the job will be built.

One problem the construction salesman has to guard against is becoming so wrapped up with a prospect that other details which should be taken care of drop out of sight. He must constantly be checking the overall selling picture. He has to be a juggler of sorts, to keep several prospects in various stages of work as well as following up with leads, and all the other nitty-gritty details he is responsible for.

Work from a prospect list. Keep it up-to-date and use it to play the percentages. Always go after the prospects with the best possibilities for success. There will be times when the list will be short and you can work with all of them. There will also be good times, and you'll be able to pick and choose.

SELL YOURSELF

The very first thing the construction salesman has to do is sell himself to Mr. X. The prospect has to have confidence in you and like doing business with you. He has to believe the salesman is going to look after his interests and protect him when necessary. You will become *his* contractor; Mr. X will call you "my" contractor just as he speaks of "my" attorney or "my" C.P.A.

Let me point out that much of the first-impression part of selling yourself will be directed toward a lead because the two of you will be meeting for the first time. The remainder of selling yourself will be directed to a prospect; it involves your personal performance while putting together the project and trying to sell it.

First Impressions.   From the very first time you meet Mr. X you start selling yourself, and you never stop. There are as many ways to sell yourself as there are people; let me give you some ways that have worked for me.

The bywords are low profile, soft sell. Everything you do is within the framework of these words.

The very first real contact Mr. X has with you is in the impression he forms when he sees you. First impressions are extremely important; therefore, I let Mr. X see exactly what he expects to see, a man who works in the construction industry. Let me ask you a couple of questions. One, how would you like to be sitting in the examining room waiting for the doctor when the door opened and, instead of a person wearing neat business clothing under a white coat with a stethoscope hanging from a pocket, in stepped someone needing a shave and dressed in jeans, mod shirt, and sandals? Two, how would you like to find your banker dressed in casual clothes for a business appointment? What would your reaction be to these two situations? I imagine the same as mine; you'd start thinking about changing doctors and bankers. Your confidence would be shaken. Now these two individuals might be competent beyond belief, but I would hesitate to give them a chance, since they didn't look the part. I'm not saying this attitude is right, only that it's human nature.

The point I want to make is that you should look like a construction person who is in management. Don't dress like you just came from between the pages of a fashion magazine. I've developed some ways to dress that I feel give Mr. X just what he expects. I wear clothes that are practical outdoors and at the same time perfectly acceptable in a place of business.

For summer my standard uniform is neat khaki pants with a short sleeve dress shirt. Depending on whom I plan to see, I may wear a tie and coat. Now, you don't have to clump around in boots either. There are many sport-type shoes that are nice-looking and yet practical—for at some time you will be on a construction site. The khaki pants and sport shoes give the construction touch, while the dress shirt, tie, and coat give you a management look—a perfect blending, I feel.

For cold weather khaki may be a little lightweight for your particular region; it is too light in Virginia in the dead of winter. Whenever possible, I wear khaki pants with dress shirt and a corduroy coat. Again the tie is optional. When the weather is bitter cold, I wear slacks, dress shirt, and a sweater plus a car coat. Now I know there are dozens of ways to accomplish our purpose: to give Mr. X the impression he is looking at a "take charge fellow" who works in the construction industry. And I'm certainly not saying for everyone to run right out and buy a new construction wardrobe. I am convinced that appearances are important, and I personally prefer to wear clothes which I feel enhance my business image.

There is one item I always carry with me—a six-inch scale in a pocket where it can be seen. Again, personal preference; it's useful, granted, but it's also seen as an accepted tool of construction management. It's sort of expected, like the law books in your attorney's office. It goes with the job, and helps Mr. X see right away a man who builds things. Admittedly several of my friends in construction, one of whom sells, kid me about my "stage prop." That's all right—everyone to his own opinion. I believe that when selling, you put out 100%.

Of course, when the occasion calls for a suit, I wear one; and many times it is simply good taste to do so. My way may not be your way of operating. The super-big construction firm will have many executives, and they'll probably look like executives. My way seems to work well for the small-to-medium construction company where management is included in all aspects of the jobs. I'll say one thing, though; I bet I'm more comfortable than the guy dressed like a banker.

Of course, if you plan to have salesmen working for you, these thoughts on the type of clothes won't affect you; but they should help in supervising your sales people.

I almost didn't include this section on appearances, for fear of seeming somewhat presumptuous—the very idea of my telling you how you should look to the prospect! But then I feel very strongly that appearances are important, and the subject should be discussed in detail.

Several of the subjects I'm going to cover now were touched on in the first chapter. I'm bringing them up again in order to have everything in this area "under one roof," so to speak.

The Handshake. Give one that is firm, brief, and friendly. I always like to lead; I put my hand out first so that there's no awkward hesitation while each party waits for the other to offer his. A psychologist probably could talk for hours about this. Many owners, especially in smaller shops where they work with their hands or supervise people who do, seem uncomfortable when shaking hands. Maybe they have trouble relating to total strangers. Be alert for this feeling, and try to put such a Mr. X at ease as quickly as possible. Remember, regardless of the shape Mr. X's hand is in, if he puts his out, you shake it!

Addressing Mr. X. I have a hard and fast rule about addressing Mr. X. He is always *Mr. X*. That is how it stays unless he instructs me to the contrary. I strongly suggest you or your salesmen follow the same rule. You can never go wrong with formal usage. If Mr. X has another form of address, then use it—for example, Dr. X or, in the case where I built the club for the police department, Lieutenant X. I've learned that older owners whom you contract with seem always to stay "Mr." Contemporary or younger owners whom you contract with seem to get around to first names as the job moves along, and I feel this is so because of the close contact necessary during construction. Even when Mr. X introduces himself as Jack X, you should forget the Jack and call him Mr. X. A little courtesy and respect will go a long way at this stage.

Why You Want to Talk With Mr. X. You, the construction salesman, are always carrying the ball. You look like a construction person when you walk in, shake hands, and introduce youself to Mr. X. What's next? Do you stand back and wait for the owner to cover you with information? Well, you'd better not. You get right down to business, and let Mr. X know that you understand he might be making plans to build a new facility. If he is, then show him you're interested in his business. That's why you took the time to come in to talk with him. Now at this point, if Mr. X is at all an aggressive operator himself, he'll appreciate someone seeking him out and asking for business. I know this from personal experience, when someone shows some initiative by calling on me and the roles are reversed—as often happens with suppliers and subcontractors wanting to work with me on negotiated work. I find myself leaning their way because they had the get-up-and-go to come to me and ask, not for the business, but for a chance to earn the business. So I see no reason why our Mr. X won't react to a construction salesman in the same manner.

Get right down to work. Find out if he's thinking about building; and if he is, then make it crystal clear that you want to know how you can help. This is your main theme, helping Mr. X. Another way to say help is to show interest. You show interest by offering to provide prices, get information on anything from drainage to construction loans, and check out any problem acting as a major roadblock to the progress of the project.

It's up to you to convey all this to Mr. X while at the same time asking questions to see if he is a lead and, if so, whether he can be qualified into a good prospect.

At first Mr. X will be on the listening end of the conversation, mainly because you know why you're there. Mr. X doesn't, and it's up to you to tell him. Bear in mind that you always do tell this within the low-profile, soft-sell guidelines, at the same time taking note of Mr. X's situation while you are meeting with him. Is he a wholesaler sitting in his office or a retailer standing behind his counter?

Continuing to Sell Yourself. All of the above takes place with a lead most of the time. Beyond this point you will have qualified the lead; and if he's a good prospect, you press on. You don't just sell yourself and do nothing else in the process. You sell yourself after the initial meeting by the way you carry out all the many steps of working up a proposal, and also all that takes place up until the contract signing. After the contract comes the construction phase; here again you keep right on selling yourself by the way you handle all the many details and problems for the owner.

The rules for accomplishing this goal of continually selling yourself are fairly simple:

1. You do what you say you're going to do.
2. You communicate with Mr. X so he knows everything that he should know.
3. You follow up.
4. You do everything you can to take the load off the prospect.

The construction salesman will have to do all four things—not two or three of them, but all four. And I'll tell you right now, it will mean a lot of hard work. The salesman who is not willing to follow these rules while working with prospects is going to be surprised at the jobs he loses. He might have a large number of leads and prospects at first but few contracts; and as contracts become fewer, so do the leads because business stimulates more business. In other words, if you don't use these rules, I don't think you'll make it as a construction salesman.

Now let's discuss each rule. First, you do what you say you're going to do. Much of what you'll be doing for Mr. X will be at your own suggestion. You'll ask if you can work up some prices, or if you can take a crack at a drainage problem; next you set out to get it done, and then report back to Mr. X. Don't leave the prospect with, "I'll see what I can do and get back to you when I can." You tell him instead, "I have a couple of ideas to check out, and I'll be back in touch in a week. Is next Friday convenient for you?" Then the following Friday see him. If you find you can make it sooner, call first and arrange a meeting. Sometimes it's impossible to get all the work done in order to call back when you're scheduled to do so. That leads us to the second rule:

Communicate with Mr. X. When you can't keep an appointment, call and tell Mr. X. While doing so, don't just say you cannot make it. Tell him why: somebody was supposed to supply you with some information and didn't do so, and you're still waiting. Or maybe something has come up that has nothing to do with the job but which is more important—then I recommend a white lie to Mr. X. For Pete's sake, don't tell him that something more important has come up! You want him to think he's the most important person you are working with.

Communication will play a vital role during the construction phase. Keeping the owner completely up-to-date lets him look good to his friends and business colleagues. When they ask why his metal building that was due to be delivered last week has been delayed, Mr. X can come back with the full story, and then go on to tell his friends he knew about the problem three weeks ago. Keeping Mr. X informed is one of your best ways of selling yourself. Keeping the prospect up-to-date allows you to accomplish another necessary function—keeping your face in front of Mr. X. It's important for the prospect to see a lot of you. It shows him you are very much interested in his business. Communicating is one of the best reasons for letting Mr. X see you.

When stopping by to report some detail to the prospect during the selling process, it's not necessary to call every time. It's perfectly okay not to phone if you're only going to be there a very short time. If Mr. X keeps you for some reason, that's fine. But if he doesn't, get in and out in a hurry. The reason why you don't want to use the phone until you're on good terms with Mr. X is that he may ask you to tell him the information while he's on the phone, thereby defeating your purpose, to be seen.

The third rule is to follow up. This rule overlaps somewhat with the second in some areas. Nevertheless it's important in itself. You don't assume anything; you check back and find out. When you assign a duty to Mr. X (remember trying to establish interest), don't wait to hear from him; check back with him. Often something will fall through a crack because each party is waiting for the other to make a move. The construction salesman has to stay in charge; so check back and follow up. This also shows the prospect interest on your part. The fact that the prospect doesn't call you when he has the information doesn't necessarily mean he is not interested; so don't wait around for Mr. X to call. In the case where he does, then super! That really expresses interest.

During the construction phase you may have to follow up behind someone who is doing work for the owner which has to be done during construction. Say the owner has some type of special equipment that goes in the slab but is not in your contract. Somebody has to take charge, and it had better be you; for that's the only way to look out for the owner's interest, and yours.

The fourth rule is that you do everything you can to take the load off the prospect. The main reason for this is to keep you in control. Except for the question used to establish the degree of interest of the prospect, you take care of answering the questions. For instance, a parking problem is noticed; right off the bat you tell Mr. X you'll check it out with city hall and get back to him. Remember to name a time for getting the details to Mr. X. When all the little nitty-gritty problems start popping up, jump right in and tell Mr. X you'll take care of this and that while he minds the store. By now you should be able to see what this does for Mr. X besides selling yourself. You're making Mr. X dependent on you. That's important!

These four rules can be boiled down into one word: dependability. The prospect has learned he can count on you, and hopefully you've dispelled a cloud I think all contractors carry with them. Take a poll on any street, asking, "What's your opinion of a contractor?" I'm willing to bet 90% of the answers will be, "They never do what they say they're going to do." Granted we are maligned unjustly most of the time. The public doesn't really understand just what kind of hardships contractors work under. Weather alone can make you into a lying bandit. The owner will think your excuse is so much baloney when you have not hit a lick on his job and the ground is still deep enough in mud to lose a small pickup, driver included, in the wink of an eye. He thinks you're doing something else profitable instead of working on his facility.

At the same time, owners have been pushed around, and they never forget it. Anyway, you have to prove to Mr. X you aren't like the rest of the bunch—you're dependable; he can count on you. And once you sell him on this, *do not* drop the ball and let it be your fault.

Return Phone Calls. Another thing you can do is return phone calls. It's amazing how many businessmen operate under the premise that if the call is important, so-and-so will call back. To me it's plain common courtesy to return a phone call. A week seldom passes that I don't recall what a prospect told me several years ago when I returned his call: "It's a pleasure doing business with a man who returns his phone calls." I might add that a 14,000-square-foot warehouse came from that particular prospect.

Now the construction salesman has to do more than return the call, be it from a lead, a prospect, or an owner during construction. The salesman has to make contact and communicate. My point is that you don't just return the call and let it drop if you miss Mr. X. You keep trying until you talk with your party. The caller will be pleased you got back to him; it also shows him you're interested in his business. All of this helps sell you to Mr. X, which is the name of the game.

Summary. Selling yourself to the prospect successfully requires all of these:

Good first impressions.
Firm handshake.
Addressing Mr. X with respect.
Quickly explaining why you want to talk to him.

and then as the selling process moves along:

Do what you say you're going to do.
Communicate.

Follow up.
Do all you can to help take the load off the prospect and make him dependent on you.

By doing your best to adhere to these rules, you'll do a good job on selling yourself to the prospect. It is hard to determine how much the prospect actually buys of you; probably the only true yardstick is the contract.

A word here about an area I stay away from—trying to sell and impress with social engagements. I don't do it, period. I feel that it's not necessary. I am a businessman doing business with another businessman; I leave it at that, and I'm positive the prospect or future owner appreciates having our dealings kept on a business level.

I'm not talking about having Mr. X join you for lunch so the two of you can get away from his phone; nor am I saying don't do it. Entertaining may be an important part of selling in your part of the country; and if you feel it helps you sell a job, then by all means, go ahead. Everyone to his own particular taste in this matter.

SELLING YOUR COMPANY

You've now sold Mr. X on yourself; the next step is to sell him on your company. In other words, you have to prove to the prospect that you can do what you say you can do. All Mr. X has from you is a lot of words, and your words have not cost him a cent so far. But can you perform? The best way to answer this question is to show the prospect past jobs and tell him some of the details as well as for whom you have worked. There are several areas to be covered; I'll list them and discuss the different aspects of each.

Pictures. We've all heard that a picture is worth a thousand words. In construction selling, it's true. Have with you at all times good top-quality pictures that will show the prospect the caliber of work your company does. Snapshots won't do; they remind me of a shoestring operation, and I believe they come across the same way to the prospect.

Snapshots can be enlarged with no trouble, but there is usually a limit beyond which the pictures start losing their clarity. The handiest size for showing details is 8″ × 10″, kept in an attractive folder. I suggest that for really good pictures you contract with a professional photographer. I've used professional pictures, and they are well worth the expense. The pro knows how to play light and shadow to highlight a particular detail or catch a building at the right time of day. The more impressive the pictures of your work are, the better job of selling you can do. It takes good tools to do a good job.

The secret to using pictures is not to overpower the prospect with them. After he's looked at eight or ten, it gets to be old-hat to him—in short, boring. Eight to ten should give a decent cross section of your work, which is all that is needed. Try to present as many different types of structures as possible.

Besides their use in selling your company, you should be aware of another use for pictures. They are great to move the job along, to get Mr. X thinking about details of his new building. This use requires more than the small thin folder used for selling the company. You need to have pictures of all different types of buildings and should break them down into categories, such as shopping centers, auto dealerships, retail stores, office buildings, storage warehouses, and so on.

These pictures don't have to be of your jobs. They can show anyone's, though you should ask permission of the owner before you take any pictures. The same is true for the jobs you have built. I don't know of anyone who has ever objected, and its a thoughtful courtesy.

Trade magazines are an excellent source of good pictures; don't hesitate to cut them out and use them. These pictures aren't shown to the prospect unless you feel they can be helpful. For example, while looking at the company pictures, Mr. X singles out one saying that's what he's thinking about, but with larger plate glass and a brick front. You know you have something close in your picture file; so you get it from the car. It doesn't have to be exactly what Mr. X has in mind, only close; but it really can give the prospect a lift to see a picture of what he wants. I sold a tire store in 1972 by using a metal building manufacturer's sales brochure that had a picture of just what the owner wanted. When I showed it to him, he said that was exactly what he wanted his new branch to look like. I had a signed contract one week later for a 7,200-square-foot retail tire store.

Pictures are a very useful sales tool, both to sell your company and the job itself; so use them.

Show Jobs. Putting the prospect in your car and showing him actual buildings your firm has constructed is another way to sell your company's ability. Be prepared for the prospect to offer some resistance to this idea. I don't know why, but it's always a big deal for Mr. X to get away for a while, even to do something that seems important. Remember, if he does take you up on being shown some jobs, it's a very good indication of interest. When you actually get Mr. X in the car, he becomes a captive audience; don't make him sorry he's in the car with you so that he can hardly wait to get back to his store. Do a little talking about the buildings you are seeing, but don't overwhelm him with construction details. He couldn't care less! He will want to hear about how much a building costs, how so-and-so is doing since he moved in, or whether that type of side-wall treatment is expensive.

Remember to do some listening; don't do all the talking. And try to show Mr. X what he's interested in; don't show him schools when he's thinking a new warehouse. Don't keep the prospect away from his business any longer than you said you would—be very careful about this. If the prospect would like to see something for the second time or a building that might be a little out of the way, then go. And don't hurry him back because you have something else to do. If I have a meeting scheduled that may go into overtime, I never commit to another appointment behind the meeting. It's very bad selling technique to have to cut Mr. X short when he's interested in talking with you. If at all possible, plan your time so there will be no conflicts.

The situation with showing jobs is the same as with pictures; they can be used to sell the job as well as the company. In showing a job you'll want to take the prospect on a tour inside if possible. And right here the bad jobs come back to haunt you—it would be something if Mr. X wanted to see the interior of the building you did last year and the job had been unpleasant at the end!

I'll never be able to say it often enough: the happy owner is the best salesman you'll ever have.

Okay, you have an owner who will let you come back; so you arrange a meeting to show his building. Always do this. Don't barge in; that's plain bad manners. When you and the prospect show up, go straight to the owner and introduce the prospect. You may be surprised; the chances are the owner will take over and conduct the tour.

Since you are in the construction business, you know from first-hand experience that no job is problem-free. Even you and the owner now conducting the tour probably had a couple of problems. The funny thing is, he probably will not mention them. He wants to give the impression that he was smart enough not to have any problems. Besides, he's flattered to be showing off his building. I'm bringing this up because you shouldn't let some little problem stop you from calling on an owner to show his building. At the same time, don't be dumb enough to ask if you know he'll refuse. Why ask for grief?

A good salesman lets his product talk for itself. You do the same thing by showing some of your past work.

Drop Names. Be a name-dropper. This is one time when it's acceptable and in good taste. Every place has its prominent citizens. They don't necessarily have to be known to the general public, say as members of city council. They can be individuals who are recognized as leaders in an area familiar to the prospect. Use their names if you have done work for them. It helps put you a step ahead of the competition when the prospect knows your firm is accepted by the "heavy hitters" in your community.

It helps as well not to hit Mr. X over the head with a name. Slip it in gently, and then let the prospect pursue the subject. For example, don't say to the prospect, "I built Mr. So-and-So's new project." That might be taken as boastful by Mr. X. What you do is say something like this, "I faced the same problem, Mr. X, just last year, and the situation was taken care of after we brought in the city engineer. It happened to be on the job we did for Mr. So-and-So." There, now, you've told the prospect your firm has done work for a certain person in hopes that it will impress him; and you've done it without being too pushy. At the same time, you're telling Mr. X you can solve problems.

Along with name dropping, be prepared to give out references. It's important here not to have a typed sheet listing all the names and phone numbers to be whipped out with a flourish. Oh, no; just offer to provide references if Mr. X wants them. He may or may not take you up on them. He might check references on his own if he takes the initiative to check owners after finding out what you've built and for whom.

If you did a job that'll hurt you, then don't show pictures, look at it, or tell who. Just hope it doesn't come up.

While giving references, offer the company bank and a name to contact there. I always offer my company lawyer as a little bonus. I feel sometime the prospect will ask his attorney to make inquiries. If you are helpful and open in this matter, the prospect's opinion of you and your firm should rise.

I also recommend using the negative reference. Now don't worry; this is not a reference who will bad-mouth you. Negative reference is a term I started using several years ago, the idea being a holdover from my earlier days on the road as a factory sales rep for a building materials manufacturer. The old hands told me when I started selling to always remember that a well-handled complaint will impress your dealer more than anything else.

I put this idea to work in selling construction. While I'm giving out the reference information, I can usually get a quick double take with, "And while you're checking, Mr. X, call Mr. So-and-So. We had the biggest headache on that job." The prospect will be a little surprised to have this type of reference offered. I go on to explain that just like him my company has its share of problems. The point I want to make with Mr. X is that what really counts is how the problems are handled. I make this very clear and invite Mr. X to call the negative reference and check for himself. I've found that such an offer carries a lot of weight and truly impresses the prospect. And the negative reference has been checked by the prospect from time to time. I believe, though, the offer is proof enough to Mr. X that you intend to shoot straight with him. It sells him on you as well as your company.

Company Brochure.   I've had occasion to use an attractive professionally done company brochure, and it can be an excellent sales tool. You're able to accomplish several things simultaneously with the company brochure. You can present the whole company image in one neat package. Pictures of past jobs along with captions giving the size and use of the buildings plus the owners' names allow you to show pictures, drop names, and provide a ready list of customer references at the same time. You accomplish all this and have the added advantage of being able to leave the material with the prospect for him to study at his leisure. It also keeps your name in front of him much better than a calling card.

Now I've had experience with company brochures, and there's one thing not to do if the pamphlet is going to be used as a sales tool: Don't show pages of employee's pictures and lists of high-powered personnel with all their qualifications. Show buildings; that's what you're selling, not people.

I've seen a brochure full of people backfire. The prospect was scared off because he saw the company was obviously carrying a very heavy overhead load and just knew it was going to be reflected in the price he paid.

To sum up, sales-oriented literature makes a fine sales tool; the company brochure that is not pitched to sales can be a hindrance rather than a help if you're not careful with its use.

Selling yourself and selling your company are really rolled up into one big selling process. It's very difficult to separate the two completely and concentrate on one at a time. Time is one problem. Not much time is required to sell your company; after Mr. X accepts the fact that your firm can indeed do the caliber of work he needs, then that's it for that subject. On the other hand, selling yourself is an ongoing process with no actual stopping point. You have to be able to jump back and forth in these two areas.

This is all fine, well, and good, you say; but how do I know when to stop this type of selling and get on with selling the job? The answer is: I don't know! Prospects are all different in some ways and alike in others. It is strictly a judgment call on the part of the construction salesman. You might call it a feeling, a change of attitude by the prospect. The successful salesman is constantly on the lookout for such changes, but don't expect them to hit you over the head. More than likely, it will be an innocent question. The best yardstick I can offer is interest on the prospect's part. Questions will be the indicator to watch for; and when you think the time is right, get to it!

SELLING THE JOB

The very first impression you want to give to the prospect is interest. You bear down on all the requirements and stress to Mr. X you need all the information possible if you are going to do a good job of helping him. "Help" is the key word. Don't start with, "Give me all the information you have." Instead start with, "In order to help get this project off the ground, Mr. X, I'll need some facts concerning first the site and then the building size." Don't throw the subject open for debate. You keep control. You ask the prospect particular questions, and you'll be pleasantly surprised how smoothly the interview goes. The prospect will be really starting to think about his new building, and the excitement will be there.

While you're working on the requirements, please remember you will not learn a thing by talking. I mention this for a special reason: As you gain more experience you'll come to know firsthand, if you don't know already, that there are certain places during the selling process where long periods of silence develop—for example, when the prospect is thinking about whether he wants a 40'-wide building or a 50' one. Your questions will make him think about the building; so for heaven's sake *let* him think. Many salesmen believe that every time a period of silence occurs they must immediately start to chatter. Fight the urge. It's hard, I know, but better to learn now than the way I did—the prospect asked me to be quiet so he could think. It was a little embarrassing; however, I still sold the job.

A heavy silence will also develop when Mr. X reads your proposal. Accept it and sit there and think about the Super Bowl, or where you're going over the weekend, or anything. Think; don't talk; let the prospect read. After the reading will come the questions, and then you can talk.

It's important not to tell the prospect what he should have. Just record his requirements and ask questions that add to the requirement knowledge. Keep in mind that you're trying to find out if the prospect has a good idea of just what it is he wants, and at the same time to work up enough data to put together a budget price. Your recommendations can come later. This method works very well with the prospect who is starting from square one.

Let's take another tack for the prospect who had his plans and was stopped because of the prices that came in when the job was put out for bid. You've got a ready-made starting point, the plans. Again ask questions; display your interest and intent to help. Using the plans, proceed to point out areas where savings can possibly be recognized. Money is the attention-getter; so sell savings and places where the cost can be reduced. Hopefully by now you've found out what the bids were and what the prospect has established for a budget. If not, then try to find out. One way is to come right out and ask. After all, the worst that can happen is that Mr. X will refuse to tell you.

You will not be refused very many times if you preface your question with this positive statement: "Mr. X, I know my firm can work up some redesign ideas to cut costs. It will be most helpful to know the budget we have to shoot for. Early on we should be able to have a good indication of how the reductions are running, and it'll save me from spinning my wheels if I know just what your budget is. I'm sure you don't want me wasting my time on a budget that will be hard to reach." You've taken a very businesslike approach, and for the most part Mr. X will respect you for it and give you the numbers you need. Now you may not actually need the figure to do your work, but this is a sure-fire method to obtain the numbers. The fact is, you'll probably really use the budget as a yardstick.

Think and Discuss Change.   This is a method I've used many times to start a job moving along seriously. Don't be afraid to suggest a radical change to Mr. X if it will help accomplish a certain goal. An example may make what I'm saying clearer. Let's say our prospect is a car dealer, and he wants the showroom and offices in one building and his service area in another, completely separate structure. You work your budgets, and the price is too high. At this point you tell the prospect the price can be reduced if he puts everything under one roof. Now this is a radical change from the original requirements, but you are showing Mr. X how to reduce costs. You're doing your job, and you're helping. He may say no; and if he does, then the prospect will have to revise his budget. Another change could be relocating the building on the site or making changes in the site itself.

This tactic is used with the square-one prospect after you've worked up a price, all the requirements have been worked out, and the price is too high. You use the radical change right off the bat with the rejected-bid prospect; it is especially effective here. In this particular case the preliminary work has been done for you.

When negotiating with a prospect who has plans, always take the plans with you and as many copies as you can get out with. Sure, you're thinking, how else will you be able to work up the needed information? Think about this one, though. What you have no one else can look at! Right? So take those plans out of circulation, and keep them out as long as you can. Usually the prospect will keep a personal set, and there's nothing you can do about it. He probably won't let them get away; so the competition still has nothing to carry away to work on.

You don't have to be a first-class construction superintendent to be able to suggest radical changes. All you're trying to accomplish at the outset is to keep the project moving. Don't even try to quote prices. Explain to the prospect that you'll have to work on the details, and you'll need the plans for a short period. Later the real construction people can tell the salesman if indeed there is a saving. However, if the salesman does have expertise in this area, then great. It will make him more comfortable with this segment of the selling process, hence making him just that much more effective.

It is imperative that the construction salesman know something about construction in general. The very first rule for a salesman is to know his product. Please remember that the radical change is an attention-getter; once in a while the prospect will buy your suggestion, but more often he will not. But you have now put yourself right smack in the middle of the project. You're now part of the picture, and that's all you wanted. Any other pluses are a bonus.

Many times a radical-change suggestion will not be necessary. Small changes may do the job just as well—substituting one type of side-wall material for a cheaper one, for instance.

Then there are the changes you recommend that will not necessarily save any money but will give the prospect a better facility for his particular use—such as a one-way mirror in his office so Mr. X can observe his sales area while working at his desk, and yet remain private. Or maybe you could work out a better traffic flow for trucks around his warehouse. All such changes sell the prospect on your interest in his project.

There are several suggestions that I've more or less standardized and fall back on when nothing else is obvious: (1) *skylights in storage areas*; (2) *windows in offices to allow Mr. X visual control* (this is a good term to become familiar with; it's one which retail and wholesale prospects will want to talk about, and by your doing so you show the prospect you know your business); (3) *plan for future expansion*; (4) the possibility that the prospect may have room for *building some space to lease out* (which I use to a lesser degree).

Somewhere in these four topics I can usually find a subject for discussion, and I've never had a prospect who didn't seem to appreciate the suggestion. With a little experience you'll be able to look at a set of plans or take the owner's requirements and see immediately how to come up with some good suggestions.

There is one special item that I always try to sell the prospect on considering, where I personally feel I would not be doing my job of looking out for the prospect's interest if I did not. The main theme is using the building as an insurance policy; I tell the prospect the advantages of designing the building to have multiple uses. Some examples will show what I mean. Mr. X wants a 60' X 100' building to operate a retail business; a 10' side-wall is more than high enough for his use. I then take the role of insurance agent and ask him what happens if something should happen to him. At first Mr. X may seem puzzled by the question, but almost always the answer is that it goes to his wife. Then I ask if Mrs. X will continue to run the business. Most of the time the answer is no; the chances are Mrs. X will sell or perhaps lease the building. Then I ask him the final question: If that be the case, then why not leave Mrs. X with something that's really worth something in the marketplace? Believe me, you'll get a quick response, and he'll ask exactly what you're talking about.

At this point, I explain to Mr. X that a 10'-high building, while working very well for his business, still limits the number of businesses which could use the structure. It would pay him to consider a 14'-high building, which would be more versatile in use, hence more marketable. I point out that he could look at the added cost as a one-payment insurance policy. Every time I bring this up, the prospect considers the suggestion; and I would say that in half the cases the owner does make the changes.

I don't like to leave the prospect thinking about not being around; so I tell him it's also a good retirement plan for him. When he does decide to retire and maybe sell or lease the building, it will be more marketable and a good source of income. This statement is always received with a smile. At the same time, I increase the contract, which makes me smile.

Often the prospect will plan to build a single-use building, and the construction salesman should stand ready to point out the pitfalls, from both a use and a financial standpoint. Bankers are a gloomy lot about lending money, and they always take the attitude that Mr. X will go under two weeks after he moves in, leaving them stuck with the building. They frown on single-purpose buildings.

There'll be times when a single-purpose building is what is called for, and nothing can be done about it. Case in point is a motor freight terminal; this particular type of building can be used for only one thing, I'm sure.

To sum up, use changes to get the prospect's attention, to impress him with your interest, and to lower already existing prices.

Budget Price. You've been hearing about the budget price all through this book. Let's now go into the subject in some detail. It should give you a better understanding of some of the other topics.

We came across the budget price in Chapter 2, and you probably remember that it was something which had to be given out with no guarantees of getting something else in return.

The budget price is just what it's called—a budget, nothing more. It's a handle for the prospect to grasp. It's necessary for the construction salesman to understand exactly what a budget is, and, even more important, to know how to use it.

There's no way to convince the prospect to move beyond a certain point until he gets a price. Everything he does is aimed in this direction. He wants to know how much the new facility is going to cost. This is even more true with a rejected-bid situation, only the main thrust is how much under the low bid can you get.

The prospect gives out his requirements only to allow you to work up the price. You're not going to get down to the nitty-gritty of selling Mr. X until you present him with a price. Then and only then will he press on—or maybe fall over in a dead faint, and forget the whole thing.

At times the budget price and the contract price are one and the same. This usually occurs with a small easy job, where from the basic requirements you're able to quote a firm contract price. This price functions the same as a budget; it gives the prospect what he has to know, and in this case he's better off because he has an exact price. In truth it doesn't eliminate the purpose of the budget, only having to work up a contract price later. As I said, this applies mainly to the small simple jobs; for the more complex ones the contract price can come only after plans and specifications are prepared for pricing.

In working up the budget price, there are many ways to save time and money. Based on your own experience with past jobs and the experience of your key subs, your estimator, if you have one, will be able to whip up a budget with no trouble. After all, that's his job—to know construction costs.

The key to a meaningful budget is a decent preliminary drawing outlining all the requirements. We went over preliminary drawings briefly in Chapter 2. A simple line sketch is all that is needed when you are talking with a prospect. As you question him and take down his requirements on your sketch, be careful to note all the details possible. This sketch need not be a super-professional drafting job. It's easy enough for the shakiest hand to make a passable drawing when grid paper is used. The grid paper ($1/8$ inch to the square is the best size, I think) looks good and allows you to work to scale. Often problems will pop up when Mr. X's ideas are put down to scale, giving the salesman a perfect opening to make a few problem-solving suggestions.

I can assure you that in making suggestions you do not have to be an architect to know what you're doing. It's really very easy; and after you've worked with several prospects, it will become even easier. With experience you should become good at it; in this area practice makes perfect. So don't be worried about blowing a contract because you think you know nothing about it; chances are you will know more than you realize.

I do recommend sitting down with subs in the plumbing, heating, air conditioning, electrical, and site work areas for a short crash course in their particular trades. It's important to know how duct work is run to service a showroom, or how the overhead lights will have to be arranged, or the most economical way to handle three bathrooms, and when you can substitute gravel for blacktop —these are the kinds of things you should know if you don't already. I've never known a sub who wasn't delighted to tell someone about his trade. Ask them; you'll be pleasantly surprised. Also when you have a special problem in a trade that takes some expertise, give it to a sub. Most of them don't get a chance to work on something really interesting very often, and they'll jump right in.

When I was building the three skating rinks for one owner, we had a lulu of a problem with the electrical system. The owner wanted all kinds of lights over the skating surface, which could be manipulated in dozens of ways to obtain certain effects while the people were skating. On top of this, everything in the entire building had to be controlled from a control center so that the manager could sit behind his console just like a disc jockey and have positive control over the rink. And the electrician had to coordinate his work with the people putting in the audio system because everything came back to the same central console.

I arranged a meeting with the electrical contractor and the owner, and the three of us sat down, told the electrician what we wanted to accomplish, and turned the project over to him. The man did a super job and really enjoyed putting the entire system together. The pride he took showed, and the rink owner received an excellent job.

My point is, use your subs because they can provide backup for your budget. I've already stated my feeling about shopping a sub after he has helped you nail a job down, but it's worth repeating: *You don't do it!*

After I have all the numbers that will go into the total budget price, I add between 5 and 7% of the total to the total amount. Now the purpose of this is to make sure you can build what the prospect wants within the price that you state.

Always explain to the prospect exactly what the budget price is. It's a not-to-exceed figure for him to use. I always stress the point that it would be very easy to "low-ball" a figure, get his name on the dotted line, and then start with the added costs. This statement accomplishes two things, I hope. It lets the prospect know I'm trying to do the best by him I can (selling myself), and puts him on notice to watch out for this maneuver from the competition.

The way the budget price is presented is just as important as the number. All prospects have some figure in mind before you give them yours. You may never know what the figure is because it may be so low, as it is in most cases, that the prospect is embarrassed to mention it. The point is, he will have a figure. It will be based on what a friend built the same kind of structure for last year, or an industry figure put out by some national headquarters, or what he built his last warehouse for ten years earlier, with his idea of rise in cost added in. Therefore, the prospect will have some idea almost every time, and chances are he'll have it broken down to cost per square foot.

You've all heard the old adage that "A little information can be dangerous." Well, it's never been more true than it is with the prospect with a square-foot price based on goodness knows what. The real hooker is that when he talks to other people, they won't always tell him the whole story—and this, I feel, is not done intentionally. The problem is that the owner confuses building cost and total construction cost. When asked what his building cost is, the owner will quote the building cost, but leave out the site work and all that goes along with it. The prospect then takes this number times his area, and, presto, he has what his new facility is going to cost in total.

Now there is another source of semifalse information for the prospect, and that is the owner who feels he paid too much, and rather than take a chance on looking foolish in the business community, will deliberately put out a lower figure. One way the embarrassed owner can answer the question and not look dumb, he feels, is to give the building cost only. In any case, the prospect still ends up with a dangerous number, which the construction salesman will have to overcome.

When the salesman presents his figure, it must be broken down into at least two segments, building cost and site cost. Never give the prospect a total turn-key price. I'm speaking from bitter experience. When I started selling construction, I worked with a turn-key price, and I was losing contracts at an alarming rate. It took me six months and the help of another salesman experiencing the same problem before we were able to solve the mystery. Our competition was quoting building cost and site work separately. Many times the site cost was never priced, only the building cost. The building cost our competition was quoting most of the time was somewhere in the ballpark the prospect was expecting, and when he compared our price to that of the competition, we were promptly forgotten.

You may ask right here, what's the big deal? Tell the prospect what your price covers, and he'll be able to make a fair assessment, you say. Well, I wish it were that easy. Let me digress a moment and tell you a not very flattering fact about prospects, and leads as well: They only hear and retain 10% of what you tell them!

When I check with someone to follow up after talking with him the month before and am told he has signed with a competitor, I don't mind it—that's business. But it really gets to me if he then says with a surprised voice, "I wish I'd known you did this kind of work." The fact is, I did tell him the month before; it just didn't register with the dummy. This to me is one of the most frustrating aspects of construction selling, one that sorely tests my objectivity. You have no way of knowing if you are really talking to the prospect or just speaking words that go in one ear and out the other. As you gain experience, you'll encounter this type of thing, and there's absolutely nothing you can do except grin and bear it.

Thus you can't explain something as important as the budget price and expect the prospect to remember how to use it. At the same time, it's confusing to the prospect to try to study the situation when he has to use two systems, especially when he knows nothing at all about construction. You cannot teach him to use your system when he has one he already understands. No, you have to make your information fit his format, not expected the prospect to learn to use something new to him.

After the other salesman and I discovered what we were doing wrong, we changed our budget price structure, and the contract-signing rate started improving immediately.

Lately I've found that building and site price have to be separate, for with all the new restrictions imposed by city hall, it's almost impossible to come close to a site price without an approved site plan from the city officials. I imagine you have found the same thing where you operate—more and more added to site work by city hall that really drives up the cost and in some cases puts the project out of sight for the owner.

When presenting the budget, push the building cost and stay away from the site details. Prospects have a way of thinking site work will cost what it will cost, and there's nothing to do but pay; so capitalize on this attitude and stress building cost and places to save. Nevertheless, chances are that the numbers you give the prospect will be more than what he has in mind. You'll know this from a long, drawn-out whistle or in some cases from a dead silence that lasts for what must seem like a year. He's trying to absorb what you've just told him; so keep quiet. He'll speak when he has his breath back.

After Mr. X has the price, you're going to have to start getting him used to the figures. Explain how costs have been and still are climbing daily. A trick I use here is to put the ball back in the prospect's court, so to speak. I ask him what prices are doing in his field, and can he buy items today for the same price that he was paying five or even ten years ago? Naturally the answer will be no, and you tell Mr. X that you can't either. The two of you are in the same boat; you just sell different merchandise.

Usually at this point it's possible to obtain the prospect's budget. Simply say, "If the price seems high, Mr. X, tell me your budget and I'll see if it's possible to reach it and stay within your requirements. I'll have to make some changes, I'm sure, but there has to be a point we can reach to move along." Once you know what he has in mind, then you can go to work on any changes that may reduce the price.

Mr. X is now waiting for you to come back so he can see if he'll be able to build or not. He is becoming dependent on the salesman. The salesman must work at having this dependency increase by handling details and problems for the prospect, every possible time.

As stated in the introduction, I'm not going to cover the numbers that go into a budget price. We are concerned with selling, not estimating. Besides every locale is different, and only you know what's going on where you operate. My only recommendation is to put in a little slush for contingencies.

There is one thing to guard against when working with the budget price, and that is the prospect who wants you to break out sub prices. Watch out! He's shopping your prices. When this comes up, I tell the prospect I'll be glad to give him a price for the structure only, and he can contract with the other trades directly. This does happen fairly frequently when you're selling metal buildings. The prospect will drop back to a structure price only mostly to save money. There is one other exception—when the prospect wants to handle a particular trade himself. Say his brother is a plumber, and he wants him to handle that area; then I will break out that price. That is the only time I will, though.

So beware of the prospect who wants a breakdown. Tell him what you will for an excuse, but don't give it to him. I handle this particular situation by telling the prospect that the prices are worked up on the past history of other jobs by the square-foot cost, it would take a great deal of time to work up sub prices, and it would not be good business to do so at this point in the negotiations. Sometimes this is a small white lie; many times it's the truth. I do use past jobs as a yardstick for a time-saving budget. But here you want to tell the prospect nicely that you'll not do it. After all, he is a prospect, and you don't want to burn any bridges.

Okay, let's talk about an area where the construction salesman does not feel very comfortable—the use of drawings with the prospect. Here's where the gray area becomes scary. Here the salesman loses some of his control.

What happens is this: You draw a sketch when recording Mr. X's requirements, work up the price, and present it to Mr. X. At the same time it's difficult to explain exactly what the price covers without using the original sketch or maybe a more detailed one you used to take sub prices. Some design-build contractors offer a simple drawing with their proposal price. I think this is a little risky, though. In any case, the prospect will want to see what the money buys; so a drawing is always part of the budget, if for no other reason than to help explain the details. It's supposed to be a simple sales aid, nothing more.

The problem arises when the prospect asks for a copy of the drawing, for any number of legitimate reasons—to show his banker or lawyer, or maybe to talk over with his spouse. Anyway, he wants a copy of the drawing.

I know you can see the trouble that might arise if you do as asked. With the drawing Mr. X has all he needs to start shopping prices. The construction salesman at this point is as vulnerable as he'll ever be. He has done a whale of a lot of work, and the prospect can take it and go on his way with it.

The only defense you have is not to leave a drawing, and sometimes when I feel the prospect might be a shopper, I don't even put the budget price in a typed letter. It's jotted on a scratch pad, and that's what I hand to Mr. X. If he tries to shop it, the competitor doesn't know if it's a price from a contractor or something that was made up. Now a price on your letterhead to proof to your competition.

Don't work under the mistaken idea that the prospect won't show your prices and drawings to a competitor. He will! I'm speaking from experience. I've had some of the most beautifully prepared proposals handed to me, without asking. When you're talking about large sums of money and the prospect sees an opportunity to shop a few thousand off the price, he's going to do it. There's not much you can do about it except do such a good selling job on the prospect that he really feels more comfortable dealing with you and will let you have the last look. For this reason the salesman never stops making the prospect dependent on him. Hopefully you will make the prospect feel a little guilty about shopping your work, and you will thereby stay in the picture.

Back to the drawing. I try to combat this situation of having to give out the drawing by using an original one while presenting my budget proposal. It doesn't have to be the one you started with; it just shouldn't be a copy. Bear with me a moment, and I'll get into the reason. Now while you're explaining your price, use the drawing to point out details and always have several questions about some particular detail. If there's a change, note it on the drawing. If no change, note that also. The main purpose is to mark up the drawing.

When you've completed your meeting and are gathering your papers, take the initiative with, "Mr. X, this is the only copy of the drawing, and we've made notes all over it; so I'm going to have it redone and brought up-to-date." I've never had a prospect argue with that approach. But nothing is 100% sure, and every now and then a prospect will ask to make a copy off his office copier for his records. For the life of me I don't know what else to say but, "Fine, go right ahead." I have tried using 12" X 20" sketch paper in hopes the large size would discourage copying. In some cases I think it does; a large number of copy machines are the slide-into variety, and there is no way to insert the 12" X 20" drawing.

There are still times when, having left a copy of the drawing, I walk away with a slightly uneasy feeling in my stomach. It comes with the territory; so get used to it. Do everything possible not to leave a drawing because you're helping the prospect become a little less dependent on you.

I know of design-build contractors who work up a full-blown proposal with simple plans of the proposed project and give them to the prospect as a matter of course. I don't recommend doing this sort of thing at all. It has to run into money, and the prospect is getting way too much for nothing.

To sum up: The budget price should not be presented as a turn-key price, all-inclusive. Break it down into building and site costs. Don't break out sub prices as a general rule. Do separate costs of building shell and trades if you feel it necessary. And don't leave drawings if at all possible.

Salesman's Exposure. This subject will overlap somewhat with the budget-price topic; so we'll jump back and forth at times. Think back to my statement that the prospect retains very little of what the salesman tells him. Well, that goes for the person of the salesman also. The only way to combat this is exposure; in other words, Mr. X has to see a lot of you.

When you are working with a normal prospect who wants to build a normal building, say a $100,000 contract, and there are no exceptional problems, the time span from first call to budget price should be around three weeks. Now I know this will often vary, but I want to give you a feel for exposure because it's important. Why take so long to come back with the budget price, you ask? A few days should be enough. That's right; a few days are enough if you only want to drop a price on Mr. X and leave to wait for his phone call. But that's not what to do to sell construction successfully. The important thing here is to use the time interval to give you exposure to the prospect.

This period between the first meeting and presenting the budget price is what I call the work-up period. During this time you should make at least four contacts with Mr. X. Two or three should be in person, and the others by phone.

It's important never to give out everything you know at the first meeting; always hold back something that you need to know. Even if you'll not be able to put the price together without the data, don't be too concerned. Preparing the price should not be very time-consuming; so you'll have time to pursue exposure and put together the price.

At the same time there usually will be questions that should be put to Mr. X; but if there are none, then you'd better get busy and think of some to ask. You'll have to be the judge of what to see Mr. X about in person and what can be handled over the phone. My guideline is, how important a question is it? To me a question is important if you cannot work up the price without its answer. Everything else, of course, is less important.

Please don't bother the prospect with obvious little questions that make no difference at all during the work-up period. They simply detract from the image the salesman is trying to establish with the prospect. Very often during the work-up period you'll have to go back to the prospect for a command decision. For example, Mr. X is undecided whether to build a 50'-wide store or one 40' wide. You offer to help by telling him if there would be any saving, and, if so, how much. Now I'm not saying for you to price two different buildings. What you provide Mr. X is just enough to make a decision, nothing more. It's really not too much of a job to determine that there would be a saving in glass. Anyway, the salesman will have to have a meeting to present the data so the prospect can make a decision or choice. Then after the building is determined, you can go on with working up the price, at the same time making sure you give yourself as much exposure as possible.

I've found that solving small problems which really don't affect the price is an excellent way to stay in contact with Mr. X during the work-up phase. For example, is the water and sewer hookup at the back or front of the site? Or, will the prospect be able to put in a 40' driveway entry without special permission from the city? All of these types of problems can usually be taken care of by phone, and then you have a very good reason to contact Mr. X with the information he's waiting for. All of this activity helps sell you to the prospect; never stop selling yourself.

Remember, it's very important to have all the exposure possible. You want the prospect to begin thinking of you as *his* contractor.

During this time the salesman can establish some rapport with Mr. X, which will go a long way when you get to the prices. It should make it a little more comfortable for all concerned.

Always have a reason to talk with the prospect. Never just stop by to pass the time of day. Mr. X doesn't have the time for this and neither should you. Have an obviously acceptable reason for calling back on him, and never, never say, "I was in the neighborhood and thought I'd stop by." No matter what the reason for calling on the prospect, you don't ever want him to feel he's being neglected in the slightest sense. You might actually be in the area, but never let Mr. X know this. You want him to think you made a special effort to talk with him about his project. Again you're selling yourself.

While it's most important for the construction salesman to gain exposure, care must be taken not to overdo it. In other words, don't become a pest and start to worry the prospect. Here is an area of selling where the salesman must exercise mature judgment. I'm at a loss to pass on any meaningful guidelines. There are so many little things that make up the prospect's personality, and the salesman is the only person who knows how to handle our Mr. X. By now you should have some idea of what you can and cannot do; so it's up to you to determine the best approach and be able to keep selling yourself without pushing his off button.

The salesman must learn to be atuned to the little things that will tell him to back off some. Maybe the prospect is a little short with you when you contact him, or it could be a put-off on a meeting. He may suddenly be hard to reach by phone—there are any number of examples. The point is that the salesman has to be able to read the signs.

A salesman friend years ago told me a short story about reading signs. He said a salesman is exactly like the Indian scout of old. Both operate mostly alone surrounded by the unfriendlies, and the only way to survive is to be able to read the danger signs left along the trail. You know, the analogy is quite true; the successful salesman is one who can note the signs, both good and bad, and can conduct himself accordingly. There's no book you can learn this art from; only studying human nature and experience will sharpen these abilities.

As I said, there are good signs as well as bad. Back in 1973 I sold a warehouse to a certain Mr. X based on one small sign from the owner's wife. I had been working with this prospect for over a month, and I was going nowhere fast. I made up my mind I'd drop the prospect after the next call if nothing developed.

Well, nothing developed—I mean of the type you can grab hold of. There was one small exception, though; the owner's wife, who ran the office, teased me in a joking manner about being a Tarheel from North Carolina. Well, up to now she had always been polite, but that's all . But now she was poking a little fun at me. I didn't quite know what to make of it until I thought about something my grandfather told me. Generally you tease somebody you like. Well-armed with that one idea, I didn't give up and eventually sold the job.

Exposure is important to selling the job, but knowing when to get it and how much are equally important. Remember, "Out of sight, out of mind."

Timing.   Timing in any sales endeavor is critical to success. It is doubly so in construction sales.

Think back to my example about the toy man who decided not to build because he was having a bad day. I'll always feel I dropped the ball with that prospect. I waited all day to take the contract to him when I should have been the first person he saw that morning after he opened his doors.

The first rule of timing is never to make a price presentation if the prospect is in a hurry. You'd better recognize the problem, and tell the prospect you'll come back when he's not so rushed. Mr. X will only hear a portion of what you have to say under the best of conditions; if his attention is elsewhere, he won't hear a thing.

Realize that when you call on the prospect he's under pressure. Pay attention when you enter his place of business. If the pressure is obvious, then get his attention and tell him you'll come back or call later. Remember you will always take second place to his business; so don't get bent out of shape when the meeting you've rearranged your schedule for is called off. Once I tried for four weeks to pick up a prospect and show him a job that he wanted to see. Every time I stopped by his place it was bedlam. I would wave and tell him I would catch him later. After a couple of times I knew the prospect was becoming embarrassed about the situation; so I went out of my way to assure him that the canceled meetings were causing me no inconvenience. He seemed to appreciate my being so understanding about his problems. And I did sell him a 4,000-square-foot building.

The second rule of timing is to try not to leave a decision to be made over the weekend. For example, when you have a price to present, and you need to get Mr. X's name on the dotted line, don't give it to him on Friday. Invariably the prospect will take it home over the weekend, and that's when the trouble starts. He has time to talk to his brother-in-law, who always knows of another contractor who he's sure can better the price. Think back to the example of the car dealer who told me on a Wednesday to put the paperwork together, after which, for reasons beyond my control, the sign-up meeting couldn't be held until the following Monday. This project fell apart over the weekend. I even hesitate to give the prospect a decision on Thursday; it has a way of ending up on Friday, then Monday.

A few years back a stockbroker friend told me about some customers who played the market daily and always sold before the weekend; their theory was that the whole world can collapse on Saturday and Sunday and affect their investments—just a point of interest to show that construction salesmen are not the only people who recognize the negative results the weekend can play on a business deal.

Momentum is as important in selling a building as it is to a football team. When the players have momentum built up, it shows in the way they march downfield rolling over the opposition. But let them stumble one time, experience a quarterback sack or an offensive penalty, and you can almost see the momentum walk off the field. The same phenomenon applies to construction selling. There is a certain driving force that the prospect gets caught up in, which makes him want to see his new facility a reality. This momentum will carry over after the contract is signed. Seldom does the prospect wonder whether he has done the right thing and try to get out of the contract. He may have passing thoughts that he keeps to himself, but once he's signed, he's committed; and he wants to keep the ball rolling.

But let the prospect get sacked by a blitzing linebacker, and he'll lose his momentum. And that's what the weekend does; it allows the prospect to sit down and think about the money, the responsibilities, the problems and, much worse, have them pointed out by someone else. The net result is a loss of momentum. Now this doesn't necessarily mean the prospect will not sign the contract. Once in a while the weekend reflection period makes the conviction to do the job that much stronger, but that's the exception. Usually the salesman will have to start all over again selling the prospect— maybe by assuring him the problems really can be solved or perhaps in the form of a change. The prospect might decide to start out with a smaller building than originally planned. In any case, the construction salesman should be prepared to "hold the prospect's hand" for a few days and work like the devil to get the momentum going again.

What happens when it's not possible to keep the prospect from having the proposal over the weekend, you ask? Nothing, is the answer. It's a dilemma that comes with the territory. The salesman's only defense is to time it so the prospect will get the proposal on a Monday or Tuesday. Now I use a tactic that will be helpful to you. I leave the final day I'll give the budget price to Mr. X tentative, which gives me a good reason to make that last exposure call. If you're thinking that's not much of a reason to call on Mr. X, just to tell him you'll see him on a later date, you're right—it's not. So I give the reason for the call substance: I tell the prospect one of my sub prices came back too high, and I'm in the process of checking it out and will be back to him next Monday. That's what I call a three-in-one shot. I timed the proposal presentation for a Monday, gave myself exposure, and let Mr. X know I'm very much interested in his project by telling him I'm doing my homework with the cost.

Even when you manage to present your proposal to the prospect at the beginning of the week, there's no guarantee you can have it returned to you before Friday. This is a difficult thing to control. Nevertheless, try to control it as much as possible, and then follow up to get the answer before the end of the week.

Watch your timing, and try to catch the prospect when he's in a receptive mood. When the prospect tells you he's ready to sign, drop everything and strike while the iron is hot. Don't delay any longer than is absolutely necessary.

Often you'll have to prepare a letter or even a contract for the prospect to sign. This happens when changes have been made to the budget price proposal and it can't be used; or it may be that the prospect never had a formal letter from you. In any case, you need something for him to sign. So you have it typed up at once. Pay overtime or anything else you have to, but do it. Then with papers in hand, go straight to Mr. X.

What you don't want to do is give the prospect the impression you're falling all over yourself to get his signature. It simply doesn't look good for the professional, so you cover the haste up with looking out for Mr. X's interest. Here's how: Usually negotiations are somewhat drawn out, and prices from subs and suppliers can become a little shaky later on. At some time during the negotiation you should have mentioned the importance of nailing down prices. Now that lays the groundwork for you to hustle and get the contract signed. You're moving fast to secure firm prices as quickly as possible, and that's why you rushed right over with the contract. If a pre-engineered metal building is involved, you have a perfect reason—to get the building in the production schedule so that the waiting period is lessened. This really is a reason when you're talking about an eight- to ten-week delivery time.

Proper timing is necessary for a smooth job after the contract is signed also. When you need a decision, try to present the situation to Mr. X when the project is on the upswing. For instance, you would like to sell the owner on a change order that will benefit him but also put a little more money in your pocket—let's say, change the windows to insulated glass, or upgrade the heating plant to take care of future expansion. Approach the owner on the upswing, when positive events are taking place, such as the slab being poured. My favorite time is when steel is going up and the building is taking shape. I think you could sell the owner almost anything when he's looking at the roof line against the sky.

Of course, you have to take into account when the work has to be accomplished and keep any changes in their proper order. But whenever possible, sell on the upswing. This also goes for bad news, which in the construction business always seems to center around time schedules.

You must move to get the prospect's name on the line, and cover your haste by telling him you're saving money and/or time. Timing is extremely critical, so learn how to make it work for you.

Helping with the Financing. As I said earlier, you'll be astounded at how little the prospect will know about money. Oh, sure, he knows everything there is about his particular field, which mainly boils down to buying and selling. The monetary mechanics of this process are not at all complicated when compared to financing a new facility.

The first problem facing the prospect is that bankers and mortgage people have their own jargon. Now, you wonder, so what? So do people in every other field of endeavor.

Well, each in his own field expects the outsider to understand what's being said to him. The problem is it's often so much gibberish to the prospect, and he's somewhat embarrassed to say he doesn't quite understand all there is to know about, e.g., take-out commitments, placement fees, or exactly what a construction loan is and how it differs from the permanent mortgage; or in some cases, if the prospect is going to be leasing space out, there may be a rent-up platform to meet. Then you can get into interest rates, time frames, and tail-end balloon payments. And there's more, much more, to utterly confuse the prospect, who only wants to build a new building so he can sell more paint, tires, or whatever.

Add to this all the paperwork required, and you think you're dealing with a government bureaucracy. In my opinion that's just about as close as you'll come to the real thing. Let me repeat here what I said earlier: if you're not comfortable with the financing aspect of selling construction, then I suggest you take the time and effort to become knowledgeable in this area. I'm sure you know a bank or mortgage company official; ask him for some help, and he'll probably flood you with reading material. It's a very good idea to stay up-to-date in this field. Learn what's new and how you can use it. Keep up with interest rates, and who's lending money at any given time.

Being able to talk over these matters will be invaluable in making yourself useful to the prospect, and that's name of the game. You must be careful, though, not to talk down to Mr. X if you see an opportunity to help with the financing. You should be as comfortable with the jargon as any banker, but don't come on like one. The idea the construction salesman wants to impart is that it's he and the prospect against the other guys.

I've learned that the best way to approach the subject with the prospect is to offer to provide him with information. It's important to give the money lenders a professional package with all the necessary data about the project at their fingertips. I touched earlier on what I call my "bank package," and now it's time to look at it in detail.

When he goes after money, the prospect is in the selling game as much as you are. He has to sell the lenders and prove at the same time he's good investment potential. One of the most effective ways to do this is for the prospect to show the loan people he's dealing with a first-rate construction firm with a track record. Money lenders are cold, hard realists; the uppermost question in their minds is, how do we pay off the loan if something should happen to Mr. X or his business? This, of course, is only one of many questions that must be satisfied, but it nevertheless is a key concern. The idea is that your bank package will show the lenders exactly what their money will be used for and put their fears to rest.

When I work up a bank package I include three parts—a cover letter explaining the requirements as given to me by Mr. X and stating the price, a description of the work to be done, and then the drawings. I always point out that the price is a good close not-to-exceed budget, but nevertheless still a budget. Then I explain that a firm contract price can't be given until architectural drawings are prepared and priced, if that is the case. In some cases the job is small, and I can give a contract price; then I state in the letter that we are dealing with a firm contract price. In all cases, architectural drawings will have to be prepared and put on file with the bank for reference during construction.

The second part is a specifications list detailing what material is to be used where. This list doesn't have to be a long super-detailed set of specifications. I hold mine to around two pages, sometimes three. Remember, we're dealing with lending people, not construction people; so the list should be kept simple and easy to read. They will not be interested in the size of the re-bar and thickness of the concrete in the foundation, only that it will be designed by a licensed architect or engineer and constructed to the design.

The third part will be two drawings. One will be a site plan showing the building location; and I don't mean a site plan designed by an engineer that you can build from, but only the size and shape, and exactly how the building is situated. The second sheet will consist of a floor plan and one or maybe two elevations. The size of the drawings is important. I know it's nice and easy to reduce them down to 8 X 11 for convenience, but I feel it helps put your image across when you give them what they're expecting: a larger-size blueprint, and I always use blueprints, not copies from the office machine. It's possible many times to have all the information put on one page, and that is convenient. But I'm after appearance as well with my bank package, and five or six pages plus two blueprints makes for a nice hefty booklet; it looks substantial. It's important not to furnish the blueprints separately but to fold and insert them in the folder cover. This keeps the package in one part, making it handier for office use.

Along with my bank package goes my offer to accompany the prospect to the bank and be on call to explain any details that may crop up. Frankly I've had to do this only twice; nevertheless, the prospect seems to appreciate knowing he can call on me.

Let me jump ahead here and answer a question I know you have. A bank package represents time and money, and you're reluctant to do all this without a contract. Well, so am I. The only time to become involved with helping the prospect secure his money is when you're protected. The way I handle this (here is where I'm getting ahead of myself) is with a contingency contract. I have a signed contract which states that if Mr. X can secure financing, I have the construction contract. I'm willing to gamble a little with the prospect, knowing that by doing all I can to help obtain the money I'm helping myself nail down a contract. I'll cover this again in Chapter 6, and go over another method to use for protection.

I do admit I don't go this far if the prospect is the least bit shaky. But by the time you get to this point in the negotiations, you should have a good readout on just what type of prospect you're dealing with.

At this stage of the negotiations, it might become necessary for you actually to look around yourself for the money. This has happened to me a couple of times; the prospect was turned down where he made application, and not because of anything he did or could not do. Often the particular lending institution in question will not make a loan for some reason. For example, many banks throw open the door if you need a million dollars, but will not even speak to you for $150,000. Many of my design-build projects fall into this $150,000 category. I've had more than one good prospect turned down by his own bank when he asked for, say, $85,000. Something like this will really throw the prospect into a tailspin. First, he's angry; then he's at a loss for the next step. That's when you step in. If you're not sure who can handle the smaller loans, then check around. In my area, which is eastern Virginia, the savings and loan associations are looking for the smaller commercial mortgages; most businessmen think they make loans only on homes and are surprised to discover otherwise.

If you are not knowledgeable here, my advice is to stay away from this area until you do feel comfortable with it. Chances are, you'll come up looking like a dummy if you don't; and one way to lose the prospect's confidence fast is for him to think you don't know what you're doing in the money field.

One word of warning when the prospect approaches the bank to apply for money seriously: Do all you can to have the job tied up because bankers think *bid*. They can put the idea in the prospect's head that maybe he should put his project out for bid. When that happens, there goes your advantage. I'm not talking about the very tentative approach the prospect uses when he has a handle on cost, to see if maybe he can go on with the project. This is usually taken care of by a phone call or a quick few minutes while making a deposit. I'm talking about the full-blown serious application for a loan. Here is where the bankers can unintentionally turn the tables on you. So watch out!

Another problem to be aware of is that it takes time to obtain a loan. Money people never seem to be in a hurry, and you have to wait for them to act. This time lag causes a loss of momentum. It's to the construction salesman's benefit to have one or two reasons to see the prospect during this period; it will help keep interest and momentum from lagging too badly.

Finally, I would like to say be prepared to lead the prospect step by step through the financial woods, and even locate a source of money if necessary. Believe me, it can mean the difference between doing and not doing the job.

Build/Lease Options. From time to time you'll come across a prospect who has all the qualifications for a profitable job except that he doesn't want to tie himself up with mortgages and the spending of his capital, which goes along with financing a project. The new facility will mean more inventory, and more inventory means he needs more money.

Usually you'll find the prospect in a dilemma as to how to go about accomplishing his goal. His problem is that he doesn't just want a larger building; he wants one that he can work from efficiently, and in a location that's convenient to his trading area. There will probably be plenty of places to rent, but most of the time he will have to take something he doesn't want and lock himself in for years on the lease. The prospect can't see himself working and buying the building for the landlord.

The prospect ends up with a special set of requirements: have a building constructed for his particular needs on the site he wants, and pay rent for a few years with the opportunity to buy the building. Who can arrange all this for him, plus work out all the details that go with a commercial project? Our prospect has a hard order to fill. He knows general contractors will build, architects will draw plans, bankers will finance, and commercial real estate people will work out the lease details; but he's in the wholesale food business and wouldn't have the time to coordinate all these people to make the project fly even if he knew how.

Then in walks the construction salesman, who explains to Mr. X that he'll take care of everything; all Mr. X has to do is sell food products. What do you think Mr. X's reaction to the salesman will be? You're right; he'll welcome him with open arms.

If you're thinking this sounds too good to be true, you're dead right! The time involved is a killer. The salesman is responsible for everything, and there won't be enough hours in the day, you'll think. The rewards will be there, though, believe me; I speak from experience: two 18,000-square-foot skating rinks, one 30,000-square-foot food warehouse, and one 20,000-square-foot moving and storage warehouse. In all of these cases the prospect wanted to keep his working capital intact and get into a new facility, and the design-build-lease plan made it possible.

This type of prospect is located and sold exactly like any other prospect with one important exception—there will be a third party involved. What used to be a two-way, one-on-one situation now becomes a building triangle of tenant, owner, and general contractor.

As I said, you follow all the guidelines with the prospect. You still need the requirements, a budget price, and as many details worked out as possible, because lease design-build will necessitate the construction salesman's selling the job twice, first to the prospect and then to the future owner. It's impossible to do so without doing your homework with the prospect. This provides the information needed to show the future owner how the numbers work out.

Along with the construction data you have to arrange for the tenant and the landlord to meet and negotiate the lease terms. Don't assume you'll be present for this particular meeting; you may not be welcome. I always offer to be in the meeting, and several times I've been present; on the other hand, I have been politely told my presence was not needed.

There is one very important point that the construction salesman must always remember: When it comes to the bottom line, you are working for the man who signs the check. This is sometimes difficult to do because you are in essence serving two masters. This arrangement at some time will lead to a disagreement; it may be just an insignificant trifle or a knock-down, drag-out. When you find yourself in this predicament, remember for whom you're working.

I'm bringing this up because I've learned that you seem to be on the prospect/tenant's side. I don't know why, except that you have built up a rapport with Mr. X, and maybe there is some empathy on your part.

After you sell the job or idea, you have to go out and locate an owner. There are two main possibilities here: First, you and/or your company can become the landlord; second, you can get one or more outsiders to be the owner.

Let's look at the first suggestion. If you or your company is in the market for some excellent real estate investments, then this type of enterprise can be very attractive. The property should appreciate with time, and hopefully there will be some cash flow as well as tax incentives. In addition to this, there is a design-build contract involved, which should provide some nice profit for the company.

There is one place to be very careful—ask for and expect complete up-to-date financial information from the prospect/tenant. When the owner commits to the tenant, he'd better be sure the tenant can meet the monthly payments.

The second suggestion takes more of the construction salesman's time; so you'd better be prepared. Now at this point "digging up" an owner is like "digging up" a lead. You have to know where to go and what to do, with one large exception, which is that you'll probably be dealing with some "heavy hitters" in the investment circles in your business community. Locating these people so you may call on them will require a combination of many methods. Of course, the personal contact is one of the best ways. The way is up to you more or less, but one thing is a must—you have to go out and ask; they're not going to knock your door down.

A good starting point when looking for an owner is to contact people in the real estate business who handle real estate developments such as shopping centers, office buildings, and warehouses. These real estate agents are always searching for people to lease space in one of their projects. If you take them a tenant, they may be able to supply an owner in return. One word of caution—security; be sure you don't put your tenant on the street to be scooped up by someone else. Until you know exactly what's going on, play your cards close to the chest.

Back in Chapter 1 I mentioned that even though professional people are not good leads, there is one way they could be invaluable to the construction salesman—as owners of design-build-lease arrangements. I'm speaking mostly of doctors, dentists, and lawyers. Their businesses generate large sums of money, and investing becomes a way of life for them. As a doctor friend told me recently, "It's what you can keep, not what you make that counts."

If you already have contacts with these people who are looking for investment opportunities, then great. If not, then I suggest you start as soon as possible because they seem to be a never ending source of owners. If one person or group is not interested, it can recommend another owner.

When you're dealing with a third-party owner, your situation is one on one, and you have a certain amount of control. It's also much easier to communicate and know where you stand. On the other hand, when you are dealing with a syndicate, it's no longer one on one, but one against a bunch. There will be one usually who brings up the question of construction cost, and asks how they know the construction costs are competitive. Believe me, this can come up. I know what you're thinking: You take the deal to them, explain the numbers that have a profit for them, and someone asks about the construction cost. You hand them a deal on a platter and someone is worried that you might be making some money out of it! Face it; there will be that particular individual to deal with. I've often wondered how such a person would react to my sending him a spec sheet for a physical exam or detailed legal work and asking for a quote. I'm mentioning this so you can cover yourself at the beginning.

Let me tell you the way I handle this situation—which works every time. As soon as I meet with the group, I tell them, in a nice way to course, that I'm there talking to them for the money I expect to make. I'm not putting in all the time and effort that a project of this caliber requires in order not to end up with the construction contract. Then I go on and explain the obvious—their main interest should be in their return. If that figure is what it should be, then what the contractor charges for the actual construction shouldn't enter the picture. If this project clicks, and we all come to terms, I want it clearly understood that I build the building! This will take care of the price-conscious man, for they are listening to you for the same reason you are there, the money. I firmly believe the investors understand, appreciate, and respect this approach. The trick here is not to handle yourself in a manner that pushes their off button. Remember, you're selling! And the sad fact may be you need them more than they need you. With the money they have, offers are coming to them all the time.

At this point the construction salesman enters a gray area where he's exposed and vulnerable. So far everything with the investors has been verbal, and to provide them with the data necessary to form an opinion, you have to disclose all the information. You're hanging on a limb, and there's precious little you can do about it except push as hard as possible to put the project together. One advantage of working with a real estate broker who understands business ventures of this sort is that he should qualify the group before you have the meeting to lay everything out for them.

Knowing they are interested ahead of your meeting will increase your chance of a sale. At least you won't be shouting from the housetops for a owner.

There is another very good source of design-build-lease ownership that should be checked out—firms which have money to invest and are looking for the chance to become the owner in a building triangle. I've personally worked with this type of firm, and their sole responsibility is to locate investment ventures and become the owner. It's a pleasure to work with them also because they are knowledgeable about the field and understand the contractor's position—especially the fact that he intends to be the one to build.

If you are not familiar with this kind of investing company, then check with people you know in the financial field—mortgage officials, bankers, and don't forget the stockbroker. They should be able to help.

There is one drawback, though: these real estate investing firms will not be interested in smaller jobs, say under $100,000. To some a $100,000 job may not be small, but I'm speaking from experience in my area. Everything is relative, I know; but I'm not able to quote figures for any area except where I work.

Let's assume the group of investors buys the package. You are now going to do exactly what I advised earlier not to do—work with a group. Please remember now that these are professional people, and that makes a difference. They are busy in their own fields, and I've found that they don't want to be involved in the daily details. Sometimes one member will be the spokesman. More often the real estate agent will act as the contact man with the contractor and the new tenant. I've found this to be a very good arrangement, mainly because you can usually get in touch with him with very little hassle.

It's important to show the outside owner or owners that you're just as much a professional as they are. Take the same pains putting together your presentation as you would a bank package. Matter of fact, the bank package can be a good outline to follow, with the investment numbers added in; always have a copy for each party who will be present.

Your biggest problem will be time. You and the prospect are ready to get the show on the road, and the group isn't able to meet for two weeks. With one owner it's not too much of a problem, but with a syndicate it seems that one is always planning to be out of town when you want to hold the meeting. It will be a trying time, and all I can say is keep swinging. If time does start to drag out, push a little to get it moving; for the longer it takes, the more apt the prospect/tenant is to start having second thoughts, and loss of momentum can set in.

The build-lease package offers a very good source of profitable business for the construction salesman if he makes the effort to learn about all the parts that constitute the building triangle. All it takes is some hard work and time. Ironically, selling the prospect is the easiest part of the triangle; locating and selling the owner or owners is the hardest part—and under no conditions should you ever forget that this owner is the party signing the checks.

Selling Parts of the Project. Whatever the reason, the construction salesman sometimes will be faced with the price being too high. The prospect cannot or will not spend the amount that is required to give him what he wants. Operating under the principle that something is better than nothing, the construction salesman has to be able to show Mr. X how to save money—and one of the best ways is to take the project apart, so to speak.

The first tactic is to suggest the prospect contract out the site work himself. Now in your budget price this is separate, but the salesman is planning to do the work. Explain to the prospect that with a site plan he can contract directly with the site contractor, and you'll be glad to supply some names. Often this is enough to eliminate the overrun. But then there will be the prospect who needs to have the price drastically reduced; when faced with this situation, I start selling the structure only. In the case of a pre-engineered metal building, it's a very simple matter to quote this part; and if you get the job, it's a fast off-and-on project. Now when you offer these options to the prospect, you must stay in control; the best way to do so is with the plans. You explain to Mr. X that along with your portion of the work, you will furnish all the necessary plans for him to contract directly with all the various subs. In addition, you'll be glad to recommend certain dependable firms for him to talk with.

My experience has been that the prospect will use the people the salesman recommends. This allows the salesman to do a good turn for the sub, which will pay dividends down the road. Of course, if the subs you recommend were the ones you were planning to use, then so much the better. They get the work, and you have a rapport with the subs that makes for better job relations during construction—and you know how important this is.

While we are discussing doing only a segment of the project, let me bring up the situation in which you might be acting in a sub capacity. In my experience, it seems to arise where metal pre-engineered buildings are involved. I never back away from a sub-position job because usually I can move on and off the project fast. At the same time, someone will have to supply plans and specifications, and I may have the chance to furnish them. Then the next step is to sell myself and try to get a larger piece of the action. What's important is to get in on the project; then try to spread out.

There is one whale of a sales angle to play when you become involved in a job as a sub. Try to impress on the owner what he stands to lose by acting as his own general contractor. Here's an excellent example of what I mean: In 1972, I contracted to build a 30,000-square-foot warehouse 30′ high for a certain Mr. X. What's interesting is the story leading up to this particular job. A couple of years before, Mr. X asked my company to quote on the metal building segment of a storage warehouse he was going to build. We quoted the job and then gave a turn-key price (building only). Mr. X opted to be his own general contractor and build the building. When the project was completed, I talked with the owner and asked him what the job had cost him. He told me around $50,000; my turn-key price was $55,000. Mr. X seemed very happy that he had saved a whole $5,000 until I dropped a bomb on him—I asked did he have any idea how much business he lost by playing general contractor, and not minding the store? I knew my question would hit home because contracting is a time-consuming business, and someone doing it on the side would have to spend a great deal of time at it. He never answered my question, but I hoped he would think about it, for I knew he was going to build again in the future. Three years later he contacted me and negotiated a turn-key price for his second warehouse. I found out later he did indeed work back, and he realized that he had probably lost money in the long run by being his own contractor.

I push this point every chance I get—every man to his particular trade; this is a good approach if you start out as a sub and hope to end up as the general contractor on the project.

Now please don't confuse taking a job apart to try to salvage something with going in as a sub and selling up. Those are two separate things.

The situations described here point up the fact that the construction salesman has to be able to think and act on his feet. Mental mobility is one of the most necessary abilities the salesman can practice. Almost every project that you're not sure about will offer you something that you can grab hold of and use to help sell the job. You just have to work at looking for the handle.

Let's return to the budget price for the moment. We discussed the changes in price you had to supply the prospect and why those changes are part of the picture. I'm bringing this subject up again to stress the point that many times these changes that necessitate new prices can and will be initiated by the salesman. When you start dismantling a budget price or selling up, you're giving the prospect some new things to think about, and cost is very much a part of this. So be prepared to do your homework and grind out the price changes, and also to offer new ideas along with the cost. This is illustrated by the case where I was asked to quote on the metal building only and I still provided a turn-key price without being asked for it.

So be prepared to really grind out price changes when dealing with these two particular situations.

Overcoming Desire to Bid Project.   Early on when you start selling construction, you'll be faced with the prospect who is not sold on negotiating the job. He wants to put it out for bid so that he knows he's getting the low price. This objection is overcome by 100% salesmanship.

The way I handle this problem is to attack rather than wait and end up on the defense. I tell the prospect at the outset all the advantages of negotiated design-build over the bid route. Now the key point here is design-build. I push the fact that the prospect is going to get exactly what he asks for designed for his particular needs. This approach appeals to his ego to a certain extent and lets him feel he's really making things happen.

The first advantage is money. Some will say this is the only advantage, and maybe they are correct. I tell the prospect if he hires an architect and bids the plans, he will supposedly know the bottom price. The fallacy here is that he only knows the bottom line for that set of plans, and those plans may call for materials which are expensive and not necessary for a functional building.

I try to hit the prospect early with the question, "Do you want a building to make money with, or do you want a monument that does the same job, only costing a small fortune?" If nothing else, I have his attention for a moment.

I keep stressing that with the design-build concept the prospect will be able to know what his planned project will cost. If he goes out to bid, Mr. X will have to spend a large sum of money to have his plans prepared; then there is no assurance the price will be in the range he wants. At the same time, the prospect has very little control over what materials are used, and this has a direct bearing on the job cost. I ask such questions as: "What do you do if the bid comes in too high and you feel it's impossible to build? And all this after spending a goodly sum with the architect. It's only natural for the architect to think for the owner that only the best will do. He's giving a good job for the money he will earn. Unfortunately often the architect will be paid by a percentage of the contract, and the more a job costs, the larger his fee."

After explaining this, I go into my sales pitch. I stress control at all steps of the project. I'll supply the prospect with necessary information as to budget prices. Then when the architect is brought in, he'll be working for the contractor, not the prospect. This working arrangement allows a give-and-take atmosphere between contractor and architect that will enable the prospect to receive the best job possible for money spent. The contractor is able to tell the architect what types of construction and materials were used in the budget price. The architect then has a set of guidelines to follow when drawing up the project. What you are eliminating basically is design creativity. The architect is paid to provide the technical expertise, but the design work has already been laid out by Mr. X. Both you and the architect are working hand-in-glove to please the prospect. The trick is to give him what he wants, and you as the general contractor can see that this happens if the architect works for you. I push the point that my firm has experience in working within a budget price and getting the most from a dollar.

The second advantage I can offer is help. Numerous problems will arise, and I will be in a position to take the load off Mr. X. I talk about helping with the finances, supplying a bank package, and being present at loan meetings. I push past performance on similar jobs (that is, if possible), and how Mr. X can profit from this experience.

The third advantage is turning to my advantage the widely held belief of the public that you can never get anything done when dealing with contractors. I hit hard on the fact that there will be one single source to communicate with. There is one overall person in charge and everything goes through him. I call this single source responsibility, and I keep driving home how it helps the prospect.

The last advantage I push is saving time. This should be important to prospects because they don't move until forced to, and then they are most likely operating with a deadline. I explain how time-consuming it is to have plans prepared and then put on the bid market. The prospect is always surprised to learn that an architect generally cannot turn around in a matter of days or even a couple of weeks. Then when I tell him it will be four to six or maybe eight weeks before he has the bid prices back in his hands, he becomes concerned. The basic problem is the prospect has no yardstick to measure performance. He knows how it's done in his business, and sees no reason why other people can't get results the way he can. Bear in mind now how the retailer, the wholesaler, and to a certain extent the industrial businessman operate. In a very broad sense, for these people to accomplish something, they simply take an item off a shelf, and send it to the customer. For many, the time it takes to crank up a construction project is incredible.

On top of this time lag within the construction industry there's the frustrating tangle of red tape associated with having site plans and drawings approved in order to secure the building permits. Now I can only speak for my working area; but every year some bureaucrat adds another regulation that makes getting the building permit more time-consuming. I'm sure you are finding the same situation where you work. And one thing is certain—it's not going to get any better.

By now the prospect is concerned about whether he's going to end up in a time crunch. At this point I put all of these negatives to work for me. I tell the prospect the architect with whom I'm associated turns out plans in a fairly short time, usually within two weeks. There are two reasons for this, which I explain to the prospect—the details are all worked out, and the architect doesn't have to provide a creative design service. He only takes the requirements and turns the information into plans to build from. Second, I'm a steady repeat client for the architect, and thus get preferred service. Now it just won't do to tell Mr. X you're faster. Talk is cheap; so explain how you're faster. After all, the prospect is no dummy; don't treat him like one. If he ever thinks you are doing that, you can forget about the contract.

Next I stress how I handle the time problem at city hall. Armed with the prospect's requirements and site information, I visit the various city departments concerned and determine exactly what they expect in the plans. This really is a time saver, since revising plans after they're turned down seems to take twice as long as it should. I especially do my homework on the site plan, for two reasons: time is the first; but the site cost is nothing but a grab in the air until you have some concrete facts to go on, and the city can provide them.

I'm selling myself to the prospect and at the same time showing him how I can get the job moving and keep it moving as well as supplying him with needed information.

I would like to point out here how I approach building officials. They're doing a job that has to be done. When I go to city hall, it's because I need help in some form. I simply ask for the help of the person I'm meeting with. Once in a while it's a sticky problem or interpreting the codes in the gray areas. I must say I've never had a building official fail to try to be as helpful as possible when I've asked for help with information in solving a problem. I know some contractors who get nothing but grief from the inspectors, but I firmly believe it's their own fault for trying to bulldoze them.

Okay, back to our subject of selling the prospect on negotiating instead of bidding. After explaining how I can save time at city hall, I start to push saving actual construction time, mainly during the start-up period. When the job is bid, each part follows the one before in a step-by-step process. Very little is done simultaneously; thus a great way of saving time is not utilized.

The example I use—since the prospect seems always to grasp it with no trouble—is buying steel. Be it a metal pre-engineered building or structural steel, there is a start-up time period; so, I explain to Mr. X, while we are waiting for the permits to clear, I put all the steel on order. This allows the order to be in the production schedule; yet it can be withdrawn before a certain time if a problem comes up and delays the project. In the case where a metal building is to be used, the building can be ordered before all the exterior details are worked out, thus saving time. Impress on the prospect that you can save him time; and if he's under the gun, that will be most important to him.

Finally I give the prospect my most important reason. I can save him money! Then I back up the statement. By knowing what construction materials cost almost on a daily basis, I'm in a position to recommend to my architect more economical ones. Also, my past experience on like jobs will allow me to know what materials will give the best service for the money spent.

I know the prospect has some idea of what he wants to pay; so I push the point that working to his budget and saving time is more important than saving a little through bidding.

Overcoming the prospect's desire to bid requires the sum total of many suggestions to him, with the emphasis on saving money and time. Quite often you will be successful; then again, if you're not, you can always bid the job and add a redesign proposal the prospect may find interesting.

Combating Competition. So far you have been in the front-line trenches of construction selling; but how do you handle yourself when the enemy overruns your position and you end up in hand-to-hand combat?

First off you'll have competition. It's the part of the free enterprise system that makes it work. Without competition we would all grow fat and lazy. Competition ensures that this doesn't happen; it keeps us sharp and up to scratch. So instead of worrying about competition, welcome it—it's the force that makes the wheels not only turn but turn faster, smoother, and for less. If you are lucky enough to have a prospect all to yourself, then I envy you. All with whom I work seem to have my competitor's literature on their desks when I walk in.

Right here I want to point out something that pertains to the previous subject. When you are selling a prospect on not bidding the job, the situation in which he's taking proposals from more than one design-build contractor can help to satisfy his intention to check on the price, and get a handle for the project from the marketplace. In short, he's sort of bidding the job. This situation goes a long way in convincing the prospect not to go out for a formal bid. It is the only case I know of where competition helps to make your job easier.

Even though the other proposals give Mr. X a good idea of what the job should cost, he's still talking to your competitors. This is your next hurdle to overcome; And there's only one way: you outsell your competition! You do this by outworking them, taking on and solving more problems, and outthinking them.

The first thing you don't do is bad-mouth your competition. A professional salesman never does this. When you do, you are in effect telling the prospect he's dumb and stupid for dealing with such a no-good so-and-so. Then guess what happens; bang! goes his off button, and you might as well pack up and leave.

My standard reply when a competitor's name comes up is that it is a good firm, and a good competitor. That's all; I'm not going to knock them, but I'm certainly not going to blow their horn either. The very first point I try to make with the prospect concerning the competition is to get him thinking about who's calling on him. For example, let's say I'm from a small- to medium-size firm, and my closest competition is a large outfit. Then I stress to Mr. X the good points about being small; I have less overhead, therefore am more competitive on price, with more personal supervision and interest in the project. See, I'm trying to have the prospect think how these positive qualities look when applied to the larger company. I want the prospect to start making comparisons between me and the others. Hopefully in this particular case he will realize the competition is indeed a large outfit, with all that overhead he will be helping to pay; and maybe he'll think about ending up as one of many and not receiving the attention he feels he should get.

This also can be used in reverse. If you're a large company, stress your good points—such as experienced workmen, volume buying of materials, and better organization for faster construction. You, as the construction salesman, will have to pitch your sales approach to bring out all the positive points. Then leave it up to Mr. X to draw his own conclusions. In my opinion the size of your prospect's business has nothing to do with stressing small or large construction firms. The small businessman may want a small contractor; then again he may like the idea of the largest outfit in town doing his work. There's really no way to tell.

Just plain hard work will overcome the competition because they may not be willing to put as much into the selling as you do. If I had to pick an area where construction salesmen, and I expect all salesmen in general, drop the ball, it would be *follow up*. I have ended up with a number of jobs not because of anything great I did; no, I just kept going back and trying to make the prospect as dependent on me as possible. You might say that I simply outlasted the competition.

So many construction companies make one or maybe two passes at a prospect then leave him to call if he wants to do business. Think back to your being on the buying end; when everything is equal—money, service, and so on—how do you choose? Most likely the same way I do; you choose the person who shows he really wants the order. So do unto your prospects as you want done to you. It's a sure way to beat the competition. But follow up always!

I've also found from experience that after the prospect receives his first prices and starts working with me, there'll be some changes which will usually add to the price, and these changes are seldom questioned; so I have a chance to improve my profit picture by a small margin.

Now let's look at what you do when you get into selling the prospect late. Your competition has already been working with Mr. X, and you're a "Johnny come lately." The very first thing you set out to do after establishing the facts pertinent to the project is to muddy the water. Your competition has a head start on selling, and you need time to catch up; so you slow everything down by giving the prospect something else to think about. Maybe an illustration will make this clear: Let's say Mr. X tells you he plans to build a 100′ × 100′ warehouse; so you work up the price, and get back to him in a few days. Timing is very important here, and I'll take this up later in detail. For now we'll plan to contact Mr. X in a few days. During this time you use the regular ways I've gone over to gain exposure while the prospect is waiting for your price. You have probably stopped the job or at least slowed it down. Then when you do present your price to Mr. X, you throw in a hooker along with your presentation. In this case you tell Mr. X you'd like for him to give some thought to maybe changing the shape of the building from 100′ × 100′ to 50′ × 200′. You explain it will save some money, and the building will be easy to divide if he ever decides to lease part or all out.

Chances are, the prospect will at least think about it. He may even call in the competition to make changes. And even if he decides against your suggestion, when the job gets back on track you're in the running. In a case where you're coming in late, pay attention when you're working on the requirements, with the idea of picking out some point that you can use to muddy the water.

There's no guarantee this tactic will work all the time. The prospect may dismiss your suggestion with a shake of his head and push on. The prospect should at this point know who you are. This at least gives you a fighting chance. Your best defense is to take the offense. Attack your competition. Do your best to impress the prospect with your special knowledge of construction. Show him a better way to get the job done. Don't be afraid to make suggestions. This is the second area where construction salesmen fall down on the job. They take the requirements as given and work from them, without a single thought on improving what Mr. X is starting with. Nothing else impresses a prospect like showing him a definite improvement, especially if it eliminates a future problem.

What you're doing here is *outthinking* your competition. I sold a job in 1974 that is a perfect example of this. Our Mr. X wanted to build a new warehouse, and one of his requirements was to be able to back his ten small delivery trucks into the building and leave them overnight. Early in the morning a crew would come in to load, or sometimes it was done at night. In any case, the trucks had to spend the night in the building. This also made the trucks secure from vandalism.

When Mr. X talked with me and my competition, he specified ten overhead doors on the back of the building and explained how he intended to use them. Well, right off the bat I could see that ten overhead doors were going to be ten headaches, and expensive ones at that. A week later I made my presentation, and received no indication at all of what Mr. X was thinking. I suspected he had heard just about the same from all involved. He could have picked any one of us out of a hat; all he had to do was make up his mind. I didn't like the odds; so I spent one full day trying to come up with something different that would improve the job in some way and make me stand out. I kept going back to all those overhead doors. That was a potential problem area, but how could I capitalize on it?

That night at home it hit me: increase the building length by 15' and put one overhead door in each corner at the back of the building, but located in the side-wall instead of the end-wall as Mr. X had requested. This would allow the trucks to be driven into the building and backed up to their loading area. So now we had two overhead doors in place of the original ten, but at the same time we had more building. Cost now became the factor to consider. The next day I was there waiting for the company estimator when he showed up at the office. He went right to work, and to my pleasant surprise and relief found that the price to Mr. X would not change. It was a wash-out because those ten doors cost a lot of money.

I called Mr. X and asked for a meeting the following day, explaining that I wanted to show him a suggestion which would solve a future problem. I then spent the rest of that day working on a drawing to illustrate my idea. I even made several scale templates to represent delivery trucks. I must admit that I puzzled Mr. X when I called him that afternoon asking for the overall size of one of his trucks. He didn't know, but found out and called me back. Needless to say, I had his interest by now.

Right on time I walked into Mr. X's office, unrolled my drawing; I told him how my idea would help him, and that it wouldn't cost any more. To prove that the new layout would work, I used the templates to show there would be ample space to jockey the trucks about. After hearing me out he lead me out back, and we marked off the approximate distances of my proposed driving area. Mr. X then had a driver move a truck around the marked-off area. There was plenty of room, and I let out a silent sigh of relief. I had the job signed up before lunch.

That's what I mean when I say out thinking your competition. There's no way I can give you a list of guidelines to follow that will enable you to do so. You'll just have to do it yourself; but if you make a point of trying, I'm sure you'll be pleased with the results.

I've often heard it said that timing is everything. I agree, it is, and very much so in construction selling. We touched on timing a couple of pages back, remember? I would like to get into what we skipped over.

The case in question is that of the timing involved in trying to slow a job down so you can catch up with your competition. After meeting with the prospect, don't rush back the next day. Drag things out if you can possibly do so, but take care not to overdo it. The reasoning here is that a construction salesman builds up momentum along with the prospect, and you want to insert a little slowdown into this momentum. If the prospect is serious, he will never really lose his, but it's amazing to me how many people selling construction grow cold on a prospect when there's a hitch. I think this goes back to my number one peeve, no follow-up.

Timing is most important when muddying the water, but it must be handled with finesse!

A good way to use timing to combat competition is to be the last salesman to make your presentation. The reason here is plain common sense. If Mr. X is taking proposals from three design-build companies, and you are first or second, there is no way you can expect to get a decision from the prospect until the third proposal is in. In a small way the third salesman controls the time when the decision will be made. The first two salesmen are being forced to conform to the third man's timetable. Always try to be the last person in. The contacts you maintain during the workup period can be most helpful with timing your proposal. It's easy during a conversation to drop some hints and find out when the competition is turning over its proposal. Ferret out such important information and use it to make sure you are last.

Along with this recommendation, keep in mind what we talked about earlier about presenting your proposal. Try to keep that time near the beginning of the week. So now you have two timing guidelines to follow: Be last, and do it early in the week. Take note too: When you're last, Mr. X can't discuss your price with the competition. It will be impossible to handle all this as you would prefer every time. You can only try, but with a little experience under your belt, it does go smoother.

My last particular comment about the competition is that just about everybody is a nice guy like you trying to make a living. As in most business areas everyone knows everybody else, and has a fair idea of what's going on; because of this it makes sense to be on pleasant terms with your competition. This can be very useful to you when selling construction. Now the one kind of prospect that the construction salesman has no defense against is the joker who is using you for a source of information and drawings if he can get them. This is the gray area where your backside is hanging in a sling, and most of the time you'll never know about the con artist until it's too late and all after the fact. I know it comes with the territory. But there's nothing that says you can't use all your available information to protect yourself. And one of the best methods of self-protection is to be on friendly terms with your competition and compare notes.

Please don't misunderstand me. I'm not advocating telling your competition everything you're doing. I'm talking about the case where it's common knowledge that you and your competition are attempting to sell a prospect, on top of which you start becoming slightly uneasy with the way things are going. You begin to smell a rat; it may be little things that are difficult to put your finger on, or it may be something like the prospect showing you a copy of another salesman's proposal and sketch. If he does it with your competitor's, then he'll do the same with yours.

When such suspicious events take place, you may decide to call your competition and compare notes. A friendly phone call may save both of you some grief. So think about it. I've done so several times, and in every case it was the right thing to do.

Don't spend your time worrying about the competition; spend it thinking about how you're going to sell the job. When you talk to a construction salesman and all he does is gripe about the competition and how they always have a lower price, then I'm willing to bet he's not much of a salesman.

Price. Please note that this last sentence was the first time I mentioned anything about a lower price. I deliberately waited until now. It's too easy to become dependent on price. When you do, you stop selling, and you're back right where you started—on the bid market.

I'm the first to say the price is extremely important; and no matter how super a salesman you are, you still won't be able to sell any price you want to put on the job. If you are mainly interested in being low, then you're not going to be much of a construction salesman. An order taker, yes, but a salesman, no!

Out of all the jobs I've sold, I would say that 75% have not been at the lowest price. Nothing else means as much to me in selling construction as hearing from the prospect, "You were a little bit higher than the other man, but I know I'm going to get a better job and that's important to me." Believe me, it's happened a number of times. When an owner tells me that, I make sure he has service like he never thought possible.

Now when the selling process deteriorates into a price war, I try to stress the important things that come with my higher price. I offer to make my price more competitive, but I tell the prospect in plain language he's not going to have this or that service. He's going to get exactly what he pays for, nothing else. Since Mr. X is a businessman, this statement will make him think. There will be times when price is the only consideration; and if the job is worth it in your estimation, then go after it. Don't get angry when the prospect insists on making a decision on price alone—it's his money.

It's important to keep in mind that the prospective projects are all different in some ways. Each one will have to be handled on an individual basis, and the construction salesman will have to be flexible in his approach. The best guideline is to do what has to be done to get the job, if you want it. So don't discount price altogether, but don't make price the only item you build your presentation around. The truly good salesman talks about price last, after he has covered all the things the price includes.

I like to bring into my proposal presentation a thought-provoking idea, and I bring it up before we get to the bottom line: *cheap and economical do not mean the same thing*. What I'm trying to do is have the prospect think about getting more for his money. I impress on him that I'm thinking economy, and not the lowest price. I want that prospect to understand that this was the guideline I used when I put together his proposal. After explaining this, I then say that if Mr. X is really thinking about a cheap building, then I'll be glad to redo the proposal with that fact in mind. Nine times out of ten the prospect tells me he wants an economical job, not a stripped-down one. I've found this to be a very good tactic; use it and I'm sure you'll see what I mean.

I think the best way to sum up this chapter on selling is to say that *service* is what you're offering the prospect. You set out at the very beginning to impress on Mr. X that you want the opportunity to work for him. Service will count because you shouldn't be offering the low price but a fair price and service. When all else is equal, chances are the prospect will decide who does the job by the service he can expect from the contractor. When you get right down to it, service is all you really have to sell—so stress it, push it, and try your best to make the prospect understand exactly what it is you can do for him.

The next point I want to emphasize is to *follow up*. Contractors have something of a bad reputation when it comes to getting things done. We are always late with the work and hard to talk to when Mr. X wants to know why we are late. Nothing ever seems to be done when the contractor says it will be. Many people do think this about contractors; so when you follow up and do what you said you were going to do when you said you would, the prospect has to be pleasantly surprised. Following up will make you more points with the prospect than anything else you can do.

I'm talking from personal experience. I can show you four buildings that I sold where the owners mentioned to me that I was the only contractor who followed up. The others didn't bother to follow up after their initial interviews, or didn't even return the owner's phone call to their offices. It's easy to see where the public gets its opinion about contractors.

Granted that such contractors are probably too busy to get involved. Okay, they should tell the prospect so, not ignore him (though on those four jobs I mentioned, I'm delighted that they did!). Follow up—it sells you and the job.

When the construction salesman enters the selling phase, it will not be a step-by-step process as I have detailed it in this chapter. Please note that you can't sell construction by the numbers. It's necessary to be familiar with all the parts that make up selling. Then you use what will obtain results at the time you feel it is right.

The salesman has to be flexible. When you are talking to a prospect, financing may come up before you have a chance to talk about your company references. The prospect may start right out wanting a ballpark figure immediately. Be prepared for anything, and let nothing upset you. The only way that prospect will know you're a pro is for you to act like one.

No matter how the selling goes, you can't go wrong by selling yourself, your company, and then the job. Then follow up.

The beautiful part about selling construction is that you know every time you do everything correctly, you keep score with contracts.

Chapter 6

Signing Up the Prospect

After working as hard as possible, you hit pay dirt when the prospect agrees to contract with you. The question you ask yourself at this point is, just how do I go about the necessary paperwork? Verbal agreements are fine in some cases but not in construction sales; it's plain good business sense when dealing in the sums of money you handle in this business to have everything on paper and signed by all parties.

The signing-up paperwork is somewhat different from and much more involved than the low-bid contract you may be familiar with. The reason is that so much of the time the salesman is working on the front end. You have a tremendous amount of work to do before you can determine the actual contract price. At the same time, you don't want to be out on a limb and spend money and time without something in writing. The answer for this interim period is a contract to contract. For want of a better word I'll use the term "letter of intent."

The letter of intent is exactly what the name implies; the prospect intends to sign a construction contract after certain conditions are met by the construction company. It gives you written protection so that time and money can be spent.

Let me point out that here you will truly begin to spend money. Right off the bat an architect will have to be called in and, most likely surveyors and site engineers. Often if a pre-engineered metal building is involved and time is a factor, the building may be placed on order. So, you see, it's absolutely necessary that you're protected by something in writing.

Now the letter of intent itself need not be a long, complicated, detailed document that tries to cover everything from A to Z; it is usually very general in nature. The details are for the contract. All the letter of intent does is allow you to work out those details.

Examples of letters of intent follow:

XYZ Food Co.
Address

City Construction Co.
Address

Re: Letter of Intent for
Construction of a New Facility

Dear Mr. Doe:

It is my intention to sign a contract with your company. It is understood by both parties this letter authorizes you to proceed with plans and specifications to determine the exact cost of the project.

It is also understood the not-to-exceed budget price is $186,500.00 and the contract price will not exceed this amount unless I make substantial changes to the plans.

Preliminary plans will be approved by me before final drawings are made, and I reserve the right to make whatever changes I deem necessary.

Sincerely,

Mr. X

February 28, 1972

Lindemann Construction Company
600 West 25th Street
Norfolk, Virginia 23571

Re: Proposed Roller Skating Rink

Attention: Mr. W. D. Booth

Dear Mr. Booth:

This letter is intended as a declaration of intent to purchase a pre-engineered metal building per our previous discussions. It is anticipated that construction will begin in the next few weeks.

It is understood by all parties that construction will be subject to site approval in either or both the Denbeigh area of Newport News and the Churchland area of Chesapeake, and further subject to suitable financial arrangements being completed.

Sincerely,

PARK PLAZA CORPORATION

Walter E. Carter

W. D. Booth

This is the type of letter I prefer to use. If you'll notice the letter written to me by Mr. X, it is on his letterhead and outlines his prerogatives as the owner. Each party knows where he stands and what is expected of him.

In my opinion, the letter carries more weight when it is on Mr. X's company letterhead. There can be no doubt he means what he's saying when he takes the trouble to write a letter instead of signing his name to the contractor's form. I know this is a small point, but it may may help; and I know it will never hurt your case if a hassle develops and a third party has to be called in.

Every time I've used a letter of intent, the prospect has asked that I compose the contents. After he checks it over and finds it suitable, the letter is typed and signed by Mr. X. I always take the original for my file.

The prospect is aware of the letter of intent before we get to that stage with his project because during the selling process I've explained how we handle the paperwork. I also make it a point to have a couple of samples to show him. This is especially good because the entire concept of design-build may be new to the prospect. He can be uncomfortable with the idea, but after seeing a letter of intent from a well-known local firm, he sees it's an acceptable way of building a new facility.

I never miss the chance to tell the prospect to ring up the company whose letter he's looking at and get a reference. Most prospects do, I might add; so use a company that will give you a good buildup.

When using a sample copy of a letter of intent, I mark out the price. I feel this is confidential information, and that's the prime reason. But at the same time you derive some help here in selling yourself. It shows the prospect you respect the owner's private company information, and therefore will handle his in the same manner.

This brings up something else the construction salesman should be on guard against—talking too much about people he's now working for or has worked for in the past. It's perfectly okay to tell for whom you have constructed buildings, but it's not okay to disclose information that should remain confidential. I mention this because sometimes a salesman will try to promote rapport with the prospect by dropping some inside information about a job in which the prospect may have some particular interest.

Discuss construction problems if you want to, but don't tell Mr. X that his competitor whom you've just finished working for had difficulties making the final payment because the two partners are thinking about splitting. Remember, your prospects are pretty astute people; one doesn't own and run a business in this day and time without being so. If you show the prospect you'll divulge private information about some other job, then you'll do it for his—and there goes the off button.

Now these same prospects will pump you for all it's worth every chance they get. For example, you might be discussing interest rates for the permanent mortgage, and the prospect knows you did a job for so-and-so six months ago; so he asks you what the rate was on that mortgage. It's none of his business; so sidestep the question. I have a method that always works; I tell the prospect I just don't remember, but why not call so-and-so and ask. This nicely puts the ball right back in the prospect's court, and he knows it.

Please be careful of this situation, and try to make yourself look good. But when you get a prospect who's a real pusher, look him in the eye and tell him, "That's confidential information, and you'll have to ask Mr. So-and-So. I'm not at liberty to talk about it." The prospect may not like being told this, but he'll trust you a little bit more.

The construction salesman will encounter all kinds of conditions that will have to be put into the letter of intent. Do anything reasonable to sign up the prospect. You have to nail down the job and get it out of the marketplace as quickly as possible. I've seen very few reasonable conditions that couldn't be worked out in some kind of acceptable manner; so put all the conditions you want to in the letter of intent. Notice I said *reasonable*. As you gain experience, you'll come to realize what is reasonable and what is downright foolish. Obtaining financing is a reasonable condition. Obtaining financing at 8% interest is not.

Even if there is more than one condition, use the letter, and don't hesitate to put all the conditions in it as long as you consider them reasonable. The prime thing is to get the prospect to make up his mind and commit to you. Once this is done, the prospect becomes the owner, and the pure selling phase is over. You are now the owner's contractor, and your main job is to get busy with clearing up or helping the owner to clear up the conditions.

The most used condition is financing, followed by rezoning and permits. Of course the one condition that will always be in the letter of intent is to bring the contract in close to the budget price. Make sure you understand exactly what this condition entails. Know that the prospect is seeking financing at the going rate, not some ridiculous figure which will be almost impossible to find. If rezoning is a condition, then do your homework and find out if you're going through an exercise in futility. Don't agree to a condition you know will likely not be met. Every once in a while you'll work with a prospect who thinks he has nothing to lose by letting you try to put the project together though he knows he has an insurmountable problem; he's more than willing to let you try. Watch out for such ringers because they can really cost you. Chances are if you do a good job qualifying the prospect, you'll know about the problems before you come to the sign-up phase. Nothing is guaranteed, though; so keep your wits about you.

The letter of intent is not worth the paper it's written on, I've had people tell me. You may be wondering the same thing about now. In my opinion the letter is worth a great deal; it is a business document signed by the prospect stating what he intends to do if certain conditions are fulfilled. What the opponents object to is that the prospect has an out. If he gets cold feet, he just tells the salesman he cannot obtain his financing, and the deal is off. They're right about this. The prospect can do just this, and there's no guarantee it won't happen.

But then what do you do if after you sign a contract it becomes known that the financing has fallen through and the owner can't pay you? What good is your contract at that point? A good friend of mine who is an attorney once told me contracts are only as good as the parties involved. If the prospect isn't going to honor his signature, then all the contracts in town won't do you one bit of good.

I prefer to think that most of the people in business are honest, and when they sign a letter stating their intentions, they expect to abide by it. If for some reason a condition is not met and there's no deal, then all the parties part friends because it was in the original agreement.

In all my experience I have had only three letters of intent that did not go to full contract. On two of them the prospects very much wanted to build. The first couldn't secure financing; the other had zoning problems that everyone thought could be worked out, including me. The last one was with a prospect who decided to back out. In all honesty, I never should have even tried to sell him and sign him up. I was uncomfortable after getting the letter of intent signed. There was nothing much I could put my finger on, just a lot of little things that didn't quite add up. I was relieved when the prospect backed out.

I don't think three busts are too bad when weighed against the many jobs I've successfully built, just as one bad egg in a dozen isn't bad at all. I'll take those odds any time.

Letters of intent are similar to competition; you can spend so much time and energy worrying about them that the job doesn't get done.

A letter of intent need not be a one-way street for the construction salesman; he can have some conditions of his own incorporated. For instance, you include the condition that if the project is not built for whatever reason, the prospect has to pay a predetermined price for the plans. This will allow you to recoup what is usually the largest expense. I'll have more to say about plans shortly.

I know the letter of intent is a very useful tool for the construction salesman. I use it all the time, and frankly I'm at a loss to understand how you can protect yourself and put the project together without one.

The letter of intent comes in all sizes and shapes. You can build it right into your proposal, and all the prospect has to do is sign on the dotted line. You can use a type of form letter where you fill in the blanks; but although this is handy, I personally don't care for it. I've found each project has its own particular conditions, and it's much easier to compose the letter to meet the situation. Matter of fact, I write almost every one of my letters of intent, and let the prospect look over my work, after which he has it typed on his letterhead. This happens because the prospect seems to have no feel for exactly what is needed.

After some exposure you'll know what works best for you. When you have determined this, then use it.

One good rule of thumb: If the time starts to stretch out after you have presented your proposal, and it's difficult to have a letter of intent signed, then watch out. Every time this happens, the job falls through! The process should not drag. You may not get the job—it may go to a competitor; but don't let the selling drag out. Selling is a percentage game, and you have nothing working in your favor when it drags. Drop it, and work on something else.

Plans and specifications will be the first order of the day after you obtain your letter. Now I'm not going to go into how you get them; suffice it to say that you operate in the construction industry so it should be no problem to have the plans prepared. We are concerned about how the owner (Mr. X is no longer a prospect, but an owner) and the drawings fit together. First of all, there will probably be a time lag while the plans are being prepared. I can only speak from personal experience about how long this lag takes, but it's usually two to three weeks, depending on the complexity of the project. I have had plans in as little as one week for a simple box warehouse.

One word of caution, though—don't let it take more than three weeks to a month at the longest. I've noticed that the momentum starts to lag after about four weeks. This shouldn't affect the contract, but it could give the owner ideas about maybe delaying the actual construction for some reason. Don't forget you're dealing with businessmen who won't see things the same way you do.

I use plans mostly to keep the enthusiasm alive during this phase. I always have the architect prepare a preliminary floor plan to a ¼" scale. Sometimes I'll include an elevation that shows storefront details if I feel it's necessary. Anyway I have these drawn as quickly as possible, and I'm back in to see the owner in about a week and a half. During that ten-day period I will have seen Mr. X at least once and maybe twice. There is always something to check with him about.

Be prepared for a very pleasant reception by the owner when you walk in with plans rolled up under your arm. This is the only time I've seen retail merchants tell someone else to take over and escort me to their private office while customers were waiting. Nothing else will turn an owner on like his first viewing of the plans. I think the plans themselves are responsible for this attitude. When the owner finally sees the first tangible proof of the reality of the project he's like a kid on Christmas morning.

Several years ago I saw an owner bend over his desk and with his arm push everything to one side when I presented his plans. Most of the paper and other items ended up on the floor. When I left an hour later, he was just sitting and looking at the blueprint. I hope you enjoy as much as I do this part of construction selling; it's plain fun to observe how an owner reacts to the blueprints for the first time.

The preliminary plans are for the owner to check, and hopefully he will make what few changes are needed. While this is being done by the owner, the site plan work is going ahead; and by the time the owner accepts the building plans, the architect should be ready to proceed.

I use of period of waiting for the final plans to have the owner mark up a floor plan for the phone company and have him confer with equipment suppliers if they have work to be done during the construction phase. For example, in the case of a retail tire facility with car lifts, preliminary work has to be put in along with the slab. The owner should be involved with all such situations. It keeps him in the picture. He knows what's happening, and the momentum is kept rolling along.

When the final plans are completed, sit down with the owner and go over them in detail. And guess what? Chances are, he won't know 50% of what he's looking at. But, nevertheless, emphasize the areas that may cause problems and hope he understands. Many times he thinks he understands but doesn't have a clear picture. Those blueprints to you are crystal clear, and construction people sometimes think everyone can read plans as well as they can. Most of the time Mr. X would not even know the bottom of the page if it weren't for the title block; so please practice patience. You may find a method I use helpful. When room sizes are discussed, I try to give the owner a visual aid. For example, if his new office is 12' × 14', I look around to find some way to show him the size. I may tell him the dimension of the room we are using as a comparison. Now size should be gone over with the first preliminary drawing and again with the final one, in hopes the owner has at least some idea of what he's getting.

Believe me, many times your best will not be good enough. You've done everything in your power to make him understand the showroom ceiling will be nine feet high, but when he sees the grid being installed, he'll tell you it's too low; he thought it was higher. You can refer back to the plans as much as you want to, and he'll always say the same thing, "I didn't understand nine feet was so low; I can't read blueprints. You're the contractor; you should have told me nine feet was so low." I've actually had this happen; so don't think it won't. Be prepared for it, and if possible put a little money in the job to cover this type of problem.

After going over the finished drawings, both you and the owner sign two sets of plans. You take one set, and Mr. X takes the other. I always file my set away, and make it a point not to let it out of my possession. This signed set of plans never gets to the job site; it stays in my office. The main reason I do this is for the protection of number one. If it ever happens that the owner tries to put one over on me, then I can haul out the plans and show Mr. X or, more importantly, a third party, just what the owner has agreed to. In all honesty I've had to do this only a couple of times, and in both cases everything blew over after the owner saw what he had agreed to. For the life of me I never did find out what the owners in question did with their signed copies of the plans. I have a hunch they were conveniently misplaced.

I said earlier that you may want to have a condition in the letter of intent that reimburses you for the cost of the plans if the job doesn't go to contract. In special cases I can see where this might help make up the prospect's mind to sign. Maybe it's some complicated type of construction that's very difficult to price until the plans are drawn. Therefore the prospect won't have a handle on price; so he has to take this route. The problem is you may find yourself providing a plans service, and there's no way to get back what it's worth for you to solve the initial problems that precede the letter of intent.

I really don't like any agreement where the prospect can buy the plans and go his merry way. I stay away from this, and do it only when I think the job is worth the calculated risk. I'm the first to say this is another of those gray areas the construction salesman encounters. The only guideline I can offer is: use common sense and know whom you're dealing with. But let me tell you right now, if you do enough construction selling, you will be taken advantage of every once in a while. It comes with the territory.

I personally know of some contractors who will prepare plans and tell the prospect to use them to take other prices. If their price is not low, then they are paid for the plans and the owner contracts with the other person. My answer to this is that I wouldn't do it under any conditions, period. You're right back on the bid market—and providing the plans as well.

The prospect can sometimes find more objections to signing than you can imagine. Most of the time this comes from a good case of being a little overcautious. You have to sign the prospect up, and at the same time give him an out. He will not be 100% sure until he has the bottom line contract price. Recognize this fact about the person you're dealing with; accept it as part of the program, and go ahead and work around it. When Mr. X knows the final contract price, then he won't hesitate to commit 100%. Don't try to force something on him that he's not comfortable with. If you do, you can blow the whole deal.

There is one other way to sign up the prospect besides the letter of intent—I call it a simple letter contract. It's very practical for smaller jobs where you can nail down cost with your sketch. I use this mostly with metal buildings when there is very little subcontract work to be done—just slab, building, and erection. It's very easy to give the prospect a firm price under these conditions. The prospect can see at once exactly what he's buying, and plans are not necessary to close the deal. Mr. X signs on the dotted line, and you're in business.

I've found from experience that the prospect is very comfortable with this type of special contract. He doesn't have to have his lawyer check over the fine print because there is none. It's clear-cut, and he knows where he stands. I even include a payment schedule so that all parties know when the money has to change hands. Now please don't try to cram too much into a letter contract. Just as soon as it loses its simplicity, its effectiveness is also lost; so keep it simple and uncluttered.

It is possible to sign up the prospect with several other contract variations that allow the prospect to get the project moving without a concrete price being determined. There can be dozens of reasons for using one of these methods.

The first is the age-old standby, cost-plus. There's no need for me to explain how it works—you're in the construction business. If you can get a cost-plus contract, then good for you. It's an ideal way to do a design-build contract. I must confess I've never been lucky enough to do a cost-plus project. Change orders and small add-ons, yes, but never from the start. I've talked with my competition about this lately, and they are not doing any cost-plus work either. It's my opinion that the great construction depression of 1974–75 made the contractor a little leaner and more competitive, giving the customers more options to take competitive prices. After all, why go cost-plus if it's no trouble to have prices supplied? At any rate there is very little cost-plus business in my working area. But it's a good way to go; take all you can get.

The other two types of sign-up contracts are actually similar. There's the contract not to exceed an established figure. I've used this type only a couple of times. The problem is that to protect yourself, you have to jack up the price. This higher price then may become an obstacle; so it takes a special situation for it. For example, once I used this type of contract because the prospect was under the gun with a deadline, and he dealt in perishable goods. For this reason he couldn't rent suitable space to move into on short notice; he had to have the new place. So in order to crank up the project at once, we agreed on this particular form of sign-up contract. I remember we went to work on the site with a line drawing of the building layout; by the time the plans were drawn and the permits issued, we had everything ready to pour the foundation. I might add that the city inspectors were very helpful and understanding. You may have a chance to use this type, but I'm sure you won't do it very often.

The next method is a variation of what I have just discussed—

contract not to exceed established price, with savings shared. I confess I haven't done a project using this idea, but I do think this type of contract has merit. It should make the prospect comfortable to know that his contractor has an incentive to try to save money. The main purpose of this contract is speed. This is the only reason, in my opinion, a prospect will move ahead without knowing the firm price of his facility.

The whole idea in any case is to get the prospect to commit to you and allow you to put the project in motion. Use whatever method will obtain results. Familiarize yourself with all the various ways to sign up the prospect, and don't hesitate to use something new you feel will do the job. Each prospect is different; you are dealing with individuals in a one-on-one relationship. Approach each one with his requirements and problems always in mind.

There is one other situation I would like to mention, that of the construction manager. I'm talking about the manager who puts the project together, plans and all, obtains prices, then steps back and lets the owner act as his own general contractor to deal with all subs directly while the construction manager acts as his adviser and inspector. He can act strictly as an adviser at the management level or use different variations of these ideas to meet the individual situation. I've never done a job this way, but I am negotiating with an owner to build his project this way right now. It really depends on the owner and what particular goal he's trying to accomplish.

The prospect is signed up for only one reason—to protect the construction salesman and the construction company. The prospect will be delighted to have you do everything and bring him the final contract with absolutely no commitment on his part until the contract is ready. Remember this. If you don't protect yourself, who will protect you?

Chapter 7

Contracts

The prospect has been sold and signed. You've solved all the problems. The firm price has been put together. It's now time to have the owner put his name on an iron-clad contract so you can start to work. At this point I have a question for you. Just what kind of contract do you have in mind? What kind of question is that, you ask? A contract is a contract; what's the big deal?

The big deal is that there is more than one type of contract. I'm not really concerned about the legalities of the contract, but how its form is accepted by the owner. From experience I know the simpler the contract, the quicker it's signed. This is most important because you don't have a job until the contract is executed. The longer this takes and the more new parties there are brought into the picture at this time, the more chances there are for delays to pop up.

I imagine most of your contract experience has been with long, very detailed documents following the A.I.A. or A.G.C. format. The contracts may have been the type those organizations furnish general contractors, or the contracts may have been prepared by the owner's lawyer in conjunction with the architect. Maybe your attorney drew up the papers. In any case, legal jargon prevails, and the contract is very likely to be more difficult to read than an insurance policy.

Another situation you may or may not have encountered when using some of the standard contracts in the construction industry is having to mark out whole sections that are not applicable to a particular situation. The owner can become uncomfortable when he sees a large legal document or when he looks at a contract with sections marked out. Mr. X is down to the wire now, and he wants to know that everything is exactly as it should be. When the owner isn't sure, he turns to his lawyer and asks that the contract be checked over. The problem here is time; an attorney is not going to peruse the contract lightly. He's going over it with a fine-tooth comb because he's paid to protect his client. This takes time, and everything stops until lawyer and client are satisfied with the contract.

When I first started selling construction, I worked for a first-rate design-build construction company that used the best short-form contract I've ever seen. Let me tell you something about that outfit and how the contract came about. The two owners were state licensed engineers, one in civil and the other in architectual engineering. The civil engineer ran the field, and the other ran the office and drew the plans. We were able to provide site and building plans from in-house personnel. This gave the sales department (two salesmen) a tremendous advantage to sell the prospects, and, believe me, it was used to the fullest extent. The operation ran smooth as silk until the contract was presented to the owner. He would take one look and head for his lawyer. Then I would spend the next month taking time that should have been used elsewhere working with the attorney until the contract suited all involved. As a salesman I didn't like having my time ground up, and the time delay made me uneasy. I never counted a job sold until the contract was signed. This situation continued for a year and was causing numerous problems. Then the sales department asked the owners to come up with a simpler, easier-to-read contract to overcome these delays. I must admit they didn't take to the idea at first; they were used to the standard forms. My colleague and I prevailed, and several months later our company lawyer worked up a contract that boiled down to three pages, with everything spelled out in plain simple English. The results were immediate— no more costly delays. (See page 175 for an example.)

I then carried it one step further. After the prospect was signed up, I'd use the contract as a reason to meet with him. I'd give him a blank copy of the contract, go over it, and suggest he let his lawyer have a look. Many times I was able to take care of a question before it became a time-stopper. Frankly, I believe that not very many copies ever showed up on a lawyer's desk. There were hardly any questions concerning protection for the owner; it was all spelled out for him in plain language. Usually details to be cleared up concerned payment and time of construction.

If you are seriously going to pursue selling construction and you do not use a shortened, easy-to-understand contract, I recommend you look into the matter. From experience I've found out it does help move the project along.

There is one negative condition the owner always seems to get around to when the contract is to be signed: a penalty clause for late completion. It comes up so often I now bring the subject up for discussion before the owner does. Let me say here I've never had to construct a design-build project where I worked under a penalty clause. I stress to the owner that the design-build job has the advantage of constant attention; if there are any delays, they will be due to matters beyond my control, such as weather, strikes, and so forth. In addition, I always tell the owner I do construction for a living; if I don't perform as I say I will, then the job starts costing me money. So it's to my personal benefit that I finish the project as quickly as possible.

Mr. X, being a businessman, will usually accept these explanations; but every now and then you'll get an owner who'll bear down on the penalty clause. Here I use a tactic that's never failed me. I tell him that since he insists on a penalty clause on completion date, and is a fair-minded businessman, I'm sure he'll be willing to agree to a bonus for early completion if I'm willing to agree to a penalty for late completion. No one yet has taken me up on my offer. I personally don't enjoy being the one doing all the giving. Try my method of handling the penalty-clause situation; I think it'll do the same job for you it's done for me.

Please don't confuse the letter contract that was discussed in the previous chapter with the short-form contract I have just gone over. Granted they are similar in that they are both abbreviated forms. The letter contract (example on page 174) is a letter that serves as an agreement to build a very simple building. I've found it useful because of the time element; that's all. If you will not feel comfortable with it, then by all means follow your preference. I personally know several contractors who use the letter contract for just about everything they build, and I mean more than simple jobs too.

I'm not trying to tell you how to run your company business, but I am telling you ideas that I've found to be very helpful in closing the deal. I've talked with people who manage salesmen, and they all tell me the same thing: one of the hardest things to teach salesmen is the art of closing the deal. It's amazing the number of good salesmen who have trouble getting the prospect's name on the contract. I don't think the word "good" should be used in this case, for if they can't close, then they certainly aren't good salesmen. When sales managers discuss personnel, the bottom line question is, "Can he close?" It's the only yardstick a salesman can be measured with. Only closers get the fame and fortune; all others should try some other occupation—there's no way they can be considered salesmen.

The worst mistake a construction salesman can make is to become complacent after he has the letter of intent. He must keep right on pushing ahead because he has not closed the sale until a contract is executed. With this goal in mind, I prefer to use the simplest contract possible that will allow me to close the deal as quickly as possible.

I have several tactics I use when presenting the contract. One is I take the approach that even when the owner obviously wants to get the project moving, he still has a little tiny cloud of doubt. Keeping this in mind I don't just hand the papers to him and wait for him to sign. As far as appearances are concerned, my main thrust is not getting the contract signed. I take the attitude that the contract is as good as signed. I bring up subjects that will require the owner's help or a decision very soon. If a metal building is involved and the color has not been chosen, I ask the owner if he can have the color by a certain time because I plan to order the building just as soon as I get back to my office. Notice I don't say just as soon as the contract is signed. The impression I want to make is that it is a foregone conclusion the contract will be signed —I am mainly interested in taking care of the loose ends.

Another good topic to bring up for discussion is the coordination between general contractor and people installing special equipment who are working directly for the owner. A good example would be a shop you plan to build that will have a bridge crane. The owner is dealing direct on the crane, but you have to provide special supports for the rails. Then when all else fails, and I really have nothing to ask the owner, I fall back on inflation. I tell the owner that his timing is perfect; with what inflation is doing to construction cost, he would have to pay a lot more for his facility had he decided to hold off. I use this only when I have nothing else to bring up—though I've found the owner will sometimes mention it while we are talking about details.

Now don't be bashful about asking for the contract. A good closer will ask for the signature at the right time. It's easy to work it right in while going over details. I say something like, "Mr. X, I brought only one copy for you to keep. If you need any more, I'll drop them off tomorrow." While talking about the contract, I suggest that we go ahead and take care of the paperwork while we are on the subject. I learned from another salesman that it's a good idea not to have signed yourself; doing so at the right moment will start the ball rolling, so to speak. I've used this suggestion successfully a number of times. Consider trying it.

Some of you may think all this to-do about how to handle the contract signing is a lot of nonsense. My answer to this is the same thing I've repeated so many times so far: selling construction is strictly percentages. You're always trying to have the percentages work for you. Consider this: In 90% of your contract closings, you are probably right; it's not necessary to go into such elaborate selling maneuvers. But for the other 10%, it very well could make the difference. The hitch is this: will you be able to tell which owner is in the 90% category and which is in the 10%? I can't! So I assume everyone is in the 10%. I believe my approach will never cost you a contract. It can do nothing but help.

Again I'm not trying to tell you how to conduct your business; I just want to make you aware of some ideas that should prove helpful when you get around to the contract signing.

No matter what form of contract you use, they all say the same thing: what you will be paid for performing a certain job. And this leads us into another item that should accompany the contract: *payment schedule.*

The payment schedule is important in a design-build project because the general contractor has already spent money to arrive at the contract stage. Also there will be anticipated expenditures to start up and man the job site. If a pre-engineered metal building is involved, more than likely a deposit will have to be sent to the manufacturer before he will fabricate the building. To cover these costs, I always ask for and get a deposit when the contract is signed or very soon afterward.

Don't just spring this deposit on the owner. After the letter of intent and before the contract is ready, I discuss how the money is to be handled and exactly what the owner can expect. My rule of thumb for the deposit is 5% of the contract amount. This 5% is flexible, though; sometimes I'll drop it to 4%, depending on the project and the owner.

I take pains to explain to the owner exactly what the deposit is for—to pay the architect, to make a deposit on a metal building, and to pay my company for time and work done to date. The owner is a businessman; and when you explain things in terms he can relate to, there's no problem. He's in business for the same reason you are—to make money.

I know you are familiar with retainages. They're something we have to put up with in the construction industry. You'll be faced with retainages in design-build as well as bid work. It's brought up by the owner, and if he insists, I put a 10% retainage clause to be applied to each invoice. I only do this now when the owner prefers it; usually when the subject comes up, I try to persuade the owner that it's not necessary on a design-build contract because he knows whom he is doing business with. It is really used for owner protection on the bid market, where the contractor is probably a complete stranger to the owner. Frankly, most of the time the owner accepts this explanation, and there is no retainage.

However, there is one thing I make clear when I do have to work with retainage being withheld. When the job is complete and the owner moves in, the full amount is due. No retainage will be owing when Mr. X enjoys the use of his new facility. Callbacks will be handled on a warranty basis, not by holding my money as a club.

This is a good place to discuss a peculiar situation that you'll run into if you do enough construction selling: that of the owner who thinks after the contract is signed and the job is coming out of the ground, he can start asking for little favors which won't cost him. Maybe you've already encountered this type of owner. In defense of some of the owners, they really have no idea what it costs to do something and think a favor or a small change won't make any difference. There's only a few bucks involved. Then you have the sharpie who's going to try to squeeze something out of you for nothing.

In the first case, if I can do what the owner asks and it involves only a few dollars, I do it. But if it's going to run into some money, I make it a point to explain, diplomatically I hope, that the change will cost some dollar amount. Then I tell him why. It could be very little material cost and high labor, or any number of variations; but the point is that I explain to the owner my situation.

I deal with the squeezer in a more blunt fashion—I put everything in relation to his business. He might be a car dealer, in which case I ask him how many extras he gives away when he sells a car. In other words, ask the owner what he gives away, and he gets the picture very quickly.

Most of the owners will have a construction loan; so it becomes necessary to provide invoices at the end of the month for the owner to turn into the bank before he makes his draw. Therefore, I make a point of explaining exactly what Mr. X can expect at the end of the month. Check with the owner's bank before the first draw is due so that you will submit an invoice that includes everything the bank needs. You certainly don't want any delays in being paid!

The payment schedule can be separate and attached to the contract, or it can be incorporated into the contract itself. The letter contract can have the payment schedule included as well (note example). It's important that all parties know how the money will be handled. Always ask for and get your front-end deposit. It's a comforting feeling to be operating on the owner's money right off the bat.

The last topic of this chapter is a situation I'm sure you as a contractor have experienced at one time or another—the case where the owner has to be protected from the big bad contractor. First, he is protected by the architect, and that is okay; it's his job; second, a performance bond is called for; and third, retainage is withheld. In other words, the contractor is held suspect, and all manner of precautions must be taken by the owner.

That's all well, fine, and good, but just who protects the contractor from the owner? The contractor agrees to build a facility costing a half million dollars, and the owner has all these checks and balances to see that he does indeed get the building. Who provides the checks and balances so the contractor knows the owner does in fact have the half million to pay him; I'll tell you who. You!

I can generally put up with some headaches from the owner if I know the money is there and waiting. The way I do this is to ask for a copy of the mortgage commitment, and confirmation from the bank making the construction loan that everything is in order. I like to do so diplomatically because I'm faced with the same problem I have when presenting the contract to the owner: 90% are going to sign; the remaining 10% may have some reservations.

The problem with getting into the owner's finances is that some businessmen are touchy about their word and their credit. Every now and then you'll encounter someone who becomes annoyed when you doubt his word by asking for proof. Most of the owners accept providing proof as the way business is done, especially when dealing in the sums of money involved. Again the problem is who's touchy and who's not. This becomes important because the situation usually comes up before the contract is signed. It should be stated in the contract that the owner will provide proof of having sufficient funds to cover the contract amount. And you certainly don't want the touchy owner to become upset at the contract signing.

I've developed a tactic that works: I become an agent for a third party who needs the information. Let me elaborate. The first thing I do is take the offensive. I point out that I have to have a copy of the mortgage commitment. Then I tell him why; my bonding company has to have the information, or I might use my suppliers. I explain that because of the large amounts of material I'll be buying from them, they want to know the bills can be paid if I, the contractor, drop out of the picture for any reason.

Now Mr. X may get a little annoyed, but not at you—rather at a third party who is just a name. You come across as a nice guy trying to do your job, and these other people have put you in the middle. Mr. X does as asked more to help you than for any other reason. Ironically, every now and then there is a case in which a third party does request this information. The metal building people really get nervous when a large order is involved if they have the least little doubt that you will be able to pay them. This doesn't necessarily mean your own honesty is in question. They may know you can't pay for the building unless the owner pays you; so they want to make sure the money is available to the contractor.

This entire chapter is for the benefit and protection of the contractor. A signed contract is the target you were shooting for way back when you started chasing down leads. It's the bottom line; so use my suggestions if you feel they will help. Along with the contract make sure the owner knows how you expect to be paid and protect yourself by knowing the owner has the money to pay you. Assume nothing; no one else will look after your interests the way you will yourself—so do it.

COMPANY LETTERHEAD

Date

Mr. K. E. Jackson
16 West Wind Avenue
Hampton, Virginia

Dear Mr. Jackson:

I propose to furnish and erect on your site the following structure:

1) One 40' × 60' × 12' pre-engineered metal building; 1:12 roof slope

2) Walls to be 26 gage galvanized, color owner's choice
   Roof to be 26 galvanized, color white

3) One 40' × 60' × 4" thick reinforced concrete slab (3000 lb) with proper footings, 6" sand fill with vapor barrier

4) Two 3070 metal walk doors with lock sets

5) One 10' × 10' steel overhead door

6) Metal building insulation to be 1½" vinyl faced fiberglass

7) Gutters and downspouts

8) Architectual plans necessary for building permit

9) Total price $22,860.00

10) Not included in price: site work, electrical work, plumbing work, HVAV work, building permit, hook-up or tap fees of any sort

11) Job time: six (6) weeks for delivery of building
    two (2) weeks to erect
    eight (8) weeks total

12) Payment schedule:

Upon securing building permit	$ 2,200.00
Upon completion of slab	5,000.00
Upon delivery of building	8,000.00
Balance upon completion	7,660.00
	$22,860.00

Sincerely,

W. D. Booth

ACCEPTED _____
                    K. E. Jackson

DATE   _____

LINDEMANN CONSTRUCTION COMPANY

Standard Form of Agreement Between Owner and Contractor

AGREEMENT

made this   12th   day of   December   in the Year Ninteen Hundred and Seventy-Two

BETWEEN

Park Plaza Corporation   the Owner, and

LINDEMANN CONSTRUCTION COMPANY, the Contractor.

The Owner and Contractor agree as set forth below.

ARTICLE I

The contractor shall perform all the Work required by the Contract Documents for the construction of

A Multi-use building to be used as a Roller Skating Center

ARTICLE II

The Work to be performed under the Contract shall be commenced within 7 days after issuance of the building permit

and completed within approximately 120 days

Contractor shall not be obligated to commence any work until the owner shall supply reasonable evidence of enforceable financial commitments in an amount sufficient to cover owner's financial obligations under this contract to the contractor, and to meet other obligations of the owner normally incurred during construction; e.g. construction loan interest, loan fees, off-site work, etc. In addition, before commencement of construction, the owner is vested with fee simple title to the real estate upon which construction is to occur.

ARTICLE III

The Owner shall pay the Contractor for the performance of the Work, subject to additions and deductions by Change Order, in current funds, the Contract Sum of

XXXXXXXXXXXXXXXXX

ARTICLE IV

Based upon Applications for Payment submitted by the Contractor to the Owner, the Owner shall make progress payments on account of the Contract Sum to the Contractor as follows:

1. Downpayment upon signing of Contract ————————— $ 7,860.00

2. Monthly progress payments ————————————— as invoiced

3. Upon delivery of Steel Components to job site ——————— 25,000.00

4. Balance upon substantial completion of the project.

Payments from owner to contractor, more than seven days late shall bear interest from the due date until paid at the rate of interest equal to 150% of the rate charged to owner by its principal source of construction funds.

ARTICLE V

The Owner shall make final payment   7   days after substantial completion of the Work.

In the event the Owner occupies the building project before substantial completion of the Work the full amount of the contract shall be due and payable at the time of occupancy.

ARTICLE VI

The Contract Documents are enumerated as follows:

1. Outline Specifications

2. Drawings 1 thru 5 titled <u>New Skating Center for Mr. Walter Carter and Park Plaza Corporation, Portsmouth, Virginia.</u>

ARTICLE VII

The Contract Documents consist of this Agreement, Supplementary and other Conditions, the Drawings, the Specifications, all Addenda issued prior to this Agreement. These form the Contract and what is required by any one shall be as binding as if required by all. The intention of the Contract Documents is to include all labor, materials, equipment necessary for the proper execution and completion of the Work.

The Contract Documents shall be signed in not less than duplicate by the Owner and the Contractor.

The term Work as used in the Contract Documents includes all labor necessary to produce the construction required by the Contract Documents, and all materials and equipment incorporated or to be incorporated in such construction.

ARTICLE VIII

The Owner shall furnish all surveys.

The Owner shall secure and pay for easements for permanent structures or permanent changes in existing facilities.

The Owner shall pay all fees and assessments levied by any municipal authority; this includes utility meter fees, water tap fees, sewer tap fees, and sewer and water line fees.

ARTICLE IX

The Contractor shall supervise and direct the Work, using his best skill and attention. The Contractor shall be solely responsible for all construction means, methods, techniques, sequences and procedures and for coordinating all portions of the Work under the Contract.

Unless otherwise specifically noted, the Contractor shall provide and pay for all labor, materials, equipment, tools, construction equipment and machinery, water, heat, utilities, transportation, and other facilities and services necessary for proper execution and completion of the Work.

The Contractor shall pay all sales, consumer, use and other similar taxes required by law and shall secure all permits and licenses necessary for the execution of the Work.

ARTICLE X

The Contractor shall purchase and maintain such insurance as will protect him from claims under workmen's compensation acts and other employee benefit acts, from claims for damages because of bodily injury, including death, and from claims for damages to property which may arise out of result from the Contractor's operations under this Contract, whether such operations be by himself or by any Subcontractor. This insurance shall be written for not less than any limits of liability required by law.

ARTICLE XI

Unless otherwise provided, the Contractor shall purchase and maintain property insurance upon the entire Work at the site to the full insurable value thereof. This insurance shall include the interests of the Owner, the Contractor, Subcontractor and Sub-subcontractors in the Work and shall insure against perils of Fire, Extended Coverage, Vandalism and Malicious Mischief.

ARTICLE XII

The Owner without invalidating the Contract may order Changes in the Work consisting of additions, deletions, or modifications, the Contract Sum and the Contract Time being adjusted accordingly. All such Changes in the Work shall be authorized by written Change Order signed by the Owner.

The Contract Sum and the Contract Time may be changed only by Change Order.

ARTICLE XIII

The Contractor shall correct any Work that fails to conform to the requirements of the Contract Documents where such failure to conform appears during the progress of the Work, and shall remedy any defects due to faulty materials, equipment of workmanship which appear within a period of one year from the Date of Substantial Completion of the Contract.

ARTICLE XIV

If the Owner fails to make payment for a period of thirty days, the Contractor may, upon seven days' written notice to the Owner terminate the Contract and recover from the Owner payment for all Work executed and for any proven loss sustained upon any materials, equipment, tools, and construction equipment and machinery, including reasonable profit and damages.

ARTICLE XV

Additional Conditions:

In the event of occupancy by Owner before substantial completion of the Work it shall be the Owner's responsibility to provide the necessary Fire and Extended Insurance Coverage in amounts at least equal to the amounts previously provided by the Contractor.

It shall be the Owner's responsibility to apply for and obtain water service, gas service, electric service, telephone service, and any other services that the Owner may desire for the operation of the Owner's building project.

1. The following amounts have been included in the Contract as allowances for anticipated work to be done:

a.	12″ pipe	$ 1,093.00
b.	15″ pipe	193.00
c.	24″ pipe	1,045.00
d.	30″ pipe	3,960.00
e.	yard drains	1,320.00
f.	man holes	1,100.00
g.	flared pipe end	175.00
h.	paint car lines	275.00
i.	car chocks	1,155.00
j.	paving	11,914.00
k.	City installed curb & gutter & drive entry	792.00
l.	site fill	4,900.00
		$27,922.00

Changes in the quantity or quality of the anticipated will be for the Owner's account.

2. Contract Conditioned Upon Owner's security permanent financing.

This agreement executed the day and year first written above.

OWNER _____   CONTRACTOR: LINDEMANN
                                                          CONSTRUCTION COMPANY

BY: _____   BY: _____

TITLE: _____   TITLE: _____

Chapter 8

Working with the Owner During Construction

With the signing of the contract you've accomplished what you set out to do—sell a design-build project. The active aggressive selling is ended; all that's left is actually building the facility. Right now if you're like 99% of the construction companies, you file away the contract, assign a superintendent, and move on to something else. And if you do, I'll guarantee that at some point during the job you'll have trouble with the owner.

Contractors have a difficult time realizing the owner in a design-build project has to have his hand held from start to finish, and I mean right up to the grand opening. I think maybe the standard attitude of the contractor comes from the manner in which he has conducted his bid business. The owner is really a stranger kept at arm's length with the plans, specifications, and contract governing every move. It's natural for the contractor to operate this way; it's the way it's always been done. From a sales standpoint, however, it's absolutely the worst way to handle a design-build owner. This is the place where the communication breaks down between the building part of the company and the salesman or salesmen, whichever the case.

If you have never sold construction before, this phenomenon will not be apparent until you've sold a job. And you employers who plan to develop a sales department, look out for some conflict between sales and the rest of the firm at this point in the job.

The thing that can cause all the trouble is the dropping of owner attention. The salesman knows how important follow-up is after the sale, for this is what referral business is built on. The salesman is looking down the road to the next prospect, and an unhappy owner is the worst thing he can have going for him.

On the other hand, the owner of the construction company has his problems too. He can't drop everything to look after one owner who might get his feelings hurt. Granted he can't stop everything, but it's still possible to give a little extra attention if the effort is made.

It's primarily the function of the construction salesman to give the attention to the job. But in this case he is dependent on the rest of the company backing him up.

I would like to discuss several subjects to give you a good idea of just what should be done in order for the owner to receive the attention he expects.

START-UP

Within a week after the contract signing and preferably just a couple of days, there should be some sign of activity on the site. The best thing to use is a job trailer, and a job sign should follow.

There can be real trouble if nothing seems to happen on the site; the owner will get upset, which many times will set the tone for the entire job. It's important for the owner to see attention given to the job.

During this period check with the owner to see if he plans a groundbreaking ceremony. If he does, then offer your help, of course; and, more important, work out the timing so the job signs are up and the top person in your construction company is there also to get his picture in the paper. This is great free publicity; take full advantage of the opportunity. Don't hesitate to bring up the topic with the owner. He just might decide to go ahead with your help, and the pictures in the paper will be worth money in your pocket. If the top man and the salesman are one and the same person, then so much the better for scheduling.

Along with these visible signs that the project has started, make every effort to get the project out of the ground. Now any good construction man knows it's to his benefit to get going, but things do have a tendency to drag sometimes. If the problem is man-power, then make sure the design-build job takes preference over a bid job, if possible. Treating the bid owner in the regular manner won't change things, but treating the design-build owner in the regular way could cause problems.

Please remember—make things happen on the site. It's when they don't that the owner gets grumpy, and the conflict between construction people and the salesman strikes sparks. The worst situation the office can put the salesman in is the one where they tell him something is going to be done and the salesman informs the owner, and then nothing happens. It makes the salesman look like all talk and no action. I don't know about you, but when I was working as a salesman, I didn't appreciate it one little bit. No one likes to appear ineffective. Always try to back your sales people; it makes life a great deal more pleasant.

WORK SCHEDULE INFORMATION

This project is the biggest thing in the owner's life right now. He not only wants but expects to be kept up-to-date on progress. He wants to know what is going to happen next. Often that's vitally important to the owner because he has to coordinate the moving of his business; the buying of new stock, maybe having new equipment delivered, and even just being able to terminate his lease all make it imperative that the owner have a good idea of the construction schedule.

I've learned that my presenting the owner with a truthful schedule is something he very much needs and appreciates. The standard bar chart used by the construction industry does the job nicely. It's easy to read and gives the owner a feel for the whole job at a glance. Just about anybody can read the bar chart, but there will be the isolated case where you may decide simply to list the schedule with the anticipated dates. Now under no condition use the critical path schedule unless you know the owner is at home with this type of graph.

I tried the critical path schedule one time. The owner was too embarrassed to admit it was still all Greek after I had explained it to him; so he went through the entire job not knowing what was going on except from what I told him. To you and me it's clear as a bell, but to that owner, it was a jumble of lines that meant nothing. Just like the blueprints, I might add.

Provide Mr. X with a work schedule he can easily understand; don't include very much detail. You want to cover the main areas, that's all.

ACQUAINTING PEOPLE ON SITE WITH OWNER

There's no way you're going to keep the owner away from the site during construction. Don't even try; instead prepare for his being there. Always let your superintendent or foreman know if possible that you and the owner or the owner by himself will be coming to the site. Make sure you introduce the owner to your key people on the site. Make sure the construction people understand Mr. X is the owner and that when they see him on the job, they are to make it a point to speak and try to be helpful. Under no condition is the owner to be ignored. It's the responsibility of the man in charge to take care of the owner when he comes on the job without the salesman.

Let me digress for a moment with some friendly advice for the construction salesman who will be working for a construction company. First realize that the bulk of the hired hands will feel just a little bit of resentment toward the salesman. There is no way you can convince them you work just as hard as, and many times harder than, they do. They see you as a good-time Joe riding around with owners showing them jobs and then maybe playing golf on the nice afternoons. You don't get your hands dirty, or freeze in the winter and bake in the summer. On top of this, these fellows know beyond a shadow of a doubt that you make as much money as the owner of the company.

Now with the workers having this opinion of you, take it nice and easy when you visit the site with or without the owner. Treat everyone as an equal; get to know the foreman if you don't already. Don't tell the construction people how to do their job or act as a spy for the owner. If something is obviously wrong, question it; but don't go running to the office. Let the foreman handle it; that's his job. I always made it a point to ask the man in charge on the job site if there were any questions I could help with. Invariably on a design-build project changes have been made, which create questions for the people driving the nails. In like manner, when a change was made during construction, I visited the site and over coffee brought the superintendent up to date. This makes his job easier and cuts down on costly mistakes. At the same time it shows the nuts-and-bolts crew you're actually involved and are working just as hard as they are at keeping the job running smoothly.

Always work with the construction people—and make sure they know who the owner is.

ACQUAINTING THE OWNER WITH BASIC PROBLEMS CONFRONTED BY THE CONTRACTOR

Explain to the owner that everything is not going to be rosy all the time. It's important that the owner be made aware of the basic problems that always seem to slow a project down. The only yard-stick the owner has is his own business, and chances are that it is an inside operation. The weather does not shut him down; and when he has a problem with his goods, he takes another item off the shelf. He is used to solving his business problems by taking immediate action.

I explain to Mr. X by relating my problems to his business. I ask what would happen if an inspector came in and, because of some fine point in the interpretation of the code for selling tires, made him shut his door until the matter was put right? How would he like the idea that every time it rained he had to stop business, his production came to a halt, but he had to continue to pay some of the help? After a few minutes of this the owner starts to have a better appreciation of what his contractor faces. It really is another world!

During the construction phase keep the owner up-to-date on the bad news if it affects the completion date. Don't let him find out from someone on the site that the metal building will be two weeks late. You'd better know yourself and then go to Mr. X and tell him, and just as important, tell him why.

At times a picture will be worth a thousand words. Take the owner to the site and show him the situation. I remember that in the fall of 1974 it seemed to rain every day for three months. We were getting further and further behind, and there was no way to catch up and make the projected completion date on a particular job. The owner was getting upset with our lack of progress. He simply could not see why, when it stopped raining, we couldn't get anything done.

I finally decided to show him exactly what we were faced with.

I took him to the site where the half-finished building stood surrounded by a sea of mud. Sitting in the mud with the tracks out of sight was a bulldozer. I told the owner that we had tried to use the bulldozer to reach the building that morning, but he could see how far we went. We even had trouble rescuing the driver. I didn't hear another word from him about our slow progress.

Communicate with the owner and let him experience the good along with the bad. Believe me, it'll help your image.

DETAILS AND "TRICKS OF THE TRADE" TO KEEP THE OWNER IN THE PICTURE

As stated earlier, the owner must not get the impression that suddenly he's no longer important. It'll be up to the salesman, or whoever sold the job (if as the owner you do your own selling), to keep Mr. X very much in the picture. I've found several methods to be helpful in doing this.

At the end of the month when the invoice is prepared, don't just mail it to him. Instead, call and ask if you can pick up Mr. X for a job visit. Tell him you have the monthly invoice, and you want to check it out with him. Now the invoice should list the work that has been completed or percent of completion if it is still in progress. The idea is to show the owner what the invoice covers. This gives Mr. X a good indication of just what progress is being made. Chances are the invoice will end up at the bank, and the bank people will inspect the site for themselves. Nevertheless, your customer is the owner, not the bank; so cater to him.

As the job progresses, small questions will come up that may or may not require a decision from the owner. Now I'm not talking about big questions; it's the small ones that can do a very nice public relations job for you. For example, the swing of an office door is in question because a vent had to be moved; or there is a question about where a vision glass should be located in the shipping foreman's office. These are the little questions that put the owner on center stage and keep him feeling that he's very much a part of the project. Even if you already know the answer, file the question away and use it when it's time to involve the owner.

Whenever possible, have the owner visit the site to make the decision; but if that is not always convenient, then you go to Mr.

X's place of business with your plans. The owner should see you at least once every ten days to two weeks. I really like to operate on a once-a-week schedule. Try not to make up trivial little things to present to the owner—he's not dumb. Use only legitimate questions, as the true professional should.

Another successful method I use is to involve the owner when it's time for interior selections to be made, such as paneling for the walls, paint colors, and floor coverings. Here you have very good reason to consult the owner. Again, don't just drop off the samples. Go over the choices, and if you know where a certain item has been used, offer to show it to Mr. X. Stay involved and be helpful at this stage.

One word of advice, though—do this well in advance of when it is needed or you may be in a pinch for time, and the owner could end up slowing the job down until final selections are made.

Don't worry too much about what you'll use for reasons to call on the owner and keep him in the picture. With a design-build project, it's amazing just how many situations come up that have to be checked out with him. The point to remember is that the owner has to be kept in the picture.

OWNER PROBLEMS DURING CONSTRUCTION

As much as you and I both would like to think that this subject is unnecessary, unfortunately that just ain't so! Take note, and believe me, you will have problems with the owner. The very best you can do is anticipate them and be prepared; hopefully staying on top of the job will allow you to catch a little problem before it blossoms into a monster. I'm now going to list what I've found to be the major problems with the owner, along with methods I use to handle the situations.

The owner thinking he's getting one thing but the plans don't agree. This problem always seems to come up, and late in the job when Mr. X can actually see exactly what he's getting. The basic problem is the inability of the owner to read blueprints. It's nobody's fault; it comes with the territory, and the construction salesman has to accept it.

I touched on this problem earlier in the book, but I feel more detail is needed. The first defense you can use is to keep drumming into Mr. X's head what the plans call for. We've talked about this; so I won't go over it again.

The second and third defenses are what I mainly want to cover. The second defense is to be astute enough when going over the plans with the owner to recognize when he's really not getting the drift of what you're explaining. He might say he does, but you know differently. This takes a little experience, but it does come with doing design-build jobs. When you sense this, or know it's a complicated area you're discussing, make a mental note. No, do better; write it down to follow up. When the area in question is being built and the shape and size can be easily seen, bring in Mr. X and go over the section so that Mr. X can see and understand what he's getting. The trick is to do this soon enough that if there are changes to be made, it won't cost too much.

When the question does pop up, you have to fall back on the plans. They show exactly what you priced and will be the basis for a change order. The owner may grumble some about assuming this and thinking that, but he'll be able to see your position.

Every now and then you'll get an owner who won't be satisfied with the way the plans read. He'll keep coming back to a difficult-to-answer statement: "That's not what I told you I wanted; never mind pointing to the plans. I know what I asked for." What do you do? Fall back to the plans and prepare for a fight? Not if you can possibly avoid it. This leads to our last defense—the slush fund, a small amount you sandbagged in the job to cover such an unpleasant situation if, when down to the wire, you felt you had to give. Don't give until you've tried every avenue of compromise, such as offering to make the change for cost, or maybe splitting the cost. In any event, for the sake of good relations, use the slush money. Cheer up; maybe the owner will ask for a legitimate change order, and you can get your money back.

A great deal of the way this unhappy situation is handled is up to you and how well you know your owner. I've only had two owners who really presented a problem, and both times I did what they wanted, simply for public relations. All the rest of the owners I've worked with who encountered this type of question couldn't have been more understanding of my situation, and in every case the problems were worked out to the acceptance of all. Sometimes

I gave more than I should, granted; but I was not forced to. There is a difference.

So now I've said my piece about this, and I hope something helps; but it will still be up to you when the owner says, "That's not what I thought I was getting."

Change Orders. I know you're familiar with how to prepare change orders: in writing, with everything spelled out and signed. But I would like to mention tolerance. Sometimes it's easy to become upset with the owner over a change order. It's a pain in the neck for job progress to be interrupted by a change order. Nevertheless, the design-build owner deserves and should receive cheerful handling of his change orders. Give him the same consideration you gave him before the contract was signed. It will all add up to recommendations in the future. Never forget how important this is.

Falling Behind Construction Schedule. No matter what the general contractor's best intentions are, the majority of projects seem to fall behind; at least all I've been associated with do. I firmly think the reason is that time is used as a selling tool. I know I've done it! You don't want the prospect to become discouraged over the excessive time span; so you present the best possible construction time frame, and hope you don't stumble. Well, somehow you do stumble, and the job starts running behind. The owner, with all his commitments, suddenly sees a crunch coming down the road, and starts raising hell!

I know you've been in this box before, but let me tell you how I handle it. Even when the job is coming out of the ground, I tell the owner every time there's a problem beyond my control that costs the job time. Then Mr. X is aware, and can judge the commitments he's made for his openings.

As the job progresses, if delays start to mount, I set up a meeting at my office with construction company management and the owner. We all sit down and go over the situation realistically. We explain our side. We also receive the owner's side, and solutions are decided upon that both sides can live with. Many times I have the job superintendent in the meeting. He's able to back up construction problems we've encountered, and at the same time gain insight into the owner's problems, from the inside so to speak. This should give him a little more incentive to try to make up some time or at least stick to the new schedule.

I might add—do your homework for a meeting of this sort because you're on the defense. Back up delay claims with proof. Order placed dates against order shipped dates are the kinds of items you need to show the owner. Have some graphs made up showing what the schedule looks like for the rest of the project, and go over them with the owner. It's imperative for you to impress on Mr. X that you're aware of his problems and are seriously doing something about the situation.

The weather, more often than not, is the main reason the job slows down. I use what I consider the ideal way to handle this with the owner. The idea comes from a company I worked for. The weather report was clipped from the paper and filed every day. At the end of the month a tally was made of the days it rained and of the clear days. The strength of the wind was also noted, since it has a direct effect on metal building erection. These weather notes will give you good back-up when meeting with the owner; so take the time to clip the information.

Now the owner has no trouble understanding bad weather as the reason, with one exception. Just as soon as the rain stops, he expects the job site to be covered with workers. One of the hardest points to get across to Mr. X is that the side effects of the rain are just as damaging as the rain itself. This is why I suggest when you explain to the owner some of your problems, you stress the side effects of rain, and how it can cripple a job. Even then you'll get phone calls from Mr. X demanding to know why there's no work being done, since, after all, it stopped raining twenty minutes ago, my advice is—be nice!

I might add that it's a good idea to have the meeting at the job site if there is something important to be seen. You can also combine the office meeting with a visit to the site. Regardless of how you put together the meeting, the important thing is to have it. And if you can strike first, your public relations image will certainly improve. What I mean is, when the job starts getting behind, call the owner and request a meeting. You are encountering some problem, and you want to bring Mr. X up to date so he will be able to plan accordingly. Don't sit around and hope the owner won't say anything. You can't hide from the problem; so attack it.

On this subject of falling behind, I've made the assumption that the delays are beyond the contractor's control. If they are within your control and the job is behind because you're not pushing it as you should, then you'll have a real problem with the owner when and if he finds out. This is the one area where the entire relationship you have so painstakingly build up with Mr. X turns into a can of worms. From this point on you're just like any other contractor, and the owner feels justified in doing anything he can to protect himself. Guess what Mr. X uses for protection? If you said money, you're correct. He starts dragging his feet on payments just as the job is dragging.

My only suggestion about this situation where the job is behind because of the contractor is—don't let it happen!

Finishing the Job and Punch List Performance.   In my opinion the last 5% is always the most difficult to get completed. The project is winding down, and hopefully you are involved with new jobs coming out of the ground. Therefore, Mr. X's job is receiving less and less of your attention. This, however, is not the way to do it.

It's extremely important for the construction salesman to work at staying on top of the finishing-up process. I know personally how hard it is when you have prospects to work, but it has to be done. I learned the hard way by having a couple of jobs turn sour in the last 5% stage.

Not only do you stay with the project, but give a little more time to the owner. It's a very trying time for him. He's facing the hassle of moving his business, plus having to coordinate with his suppliers. Some personal attention and help with getting the building ready will go a long way when referral times comes around.

When the project is very near completion, arrange for a walk-through meeting with the owner. If you have a boss, make sure he's there, as well as the foreman responsible for seeing that the work is completed. Give yourself plenty of time. These punch list walk-throughs can run into time because the owner will be enjoying his new building.

Now after the punch list is worked up, get the work done, all of it, in a reasonable time frame. Arrange a second meeting to go over the completed punch list. Usually the salesman by himself can take care of this. Please check the job ahead of time so you won't be surprised and end up with egg on your face.

Remember, get the project completed.

Finally Mr. X is in his new building, but the salesman's work is not done yet. Two other matters should be attended to.

The first is the owner's grand opening. If there is one, be there; and if you have a boss, have him there also. Mr. X is pleased as can be with his new facility and wants very much to show it off to all comers. As close as he's worked with the contractor during the job, he'll feel slighted if the contractor doesn't take the time to attend and help him enjoy the occasion. There's another reason to show up: it's a great place to pick up leads; so mingle with the crowd! There's a good chance some of Mr. X's competitors will stop by to see the new facility, enabling you to do a little self-serving lead gathering.

The second chore left for the salesman is to follow up in about three months with the owner to check on any problems. Now most owners have never known of a contractor to check back. This follow-up call really impresses the owner. Matter of fact, I've picked up several good referral leads from an owner while making this type of call. You may wonder why the owner didn't call you about the lead; but probably he just forgot it until he saw you.

Also there will be callbacks right after the owner occupies the building. See that they are taken care of, and then personally check back. It's this kind of service that reputations are built on; it means business in the future.

The construction salesman who sees the owner through all the trials and tribulations of construction will be making sure that owner becomes a ready source of business for him. The happy, smiling owner will be your best salesman. This happy result doesn't come free; you have to work for it, but the rewards are well worth it.

Chapter 9

The Construction Sales Department

This chapter is intended primarily for the construction executive who is interested in forming a sales department. When I mention sales department, it can mean one or more salesmen. If you have or are planning to employ one salesman, and his responsibilities will be selling, then you have a sales department.

The idea of a sales department may take some getting used to, just like the topics in Chapter 2. But it's my understanding that some of the larger construction companies in the United States have salesmen and have used them for several years. It's not a new idea, and it is being used within the construction industry successfully.

I have no doubt whatsoever that more and more general contractors will be using professional salesmen in the future. As time passes, we will experience a great many changes throughout the industry, and construction selling will be a part of these changes. I hope my thoughts and the experiences that I relate in this chapter will help you form and operate a successful sales department.

The first thing I should explain is the type of person you'll employ as a salesman. Again, it's probably going to take some getting used to.

You'll have someone working for you who to all appearances will look like he's not. He'll wander in and out of the office on no set schedule. Most of the time you'll have a vague idea about where he is and what he's doing. Some of the time you won't have the foggiest notion of where he is. You may go for a couple of days without seeing him. You call his home in the middle of the morning to leave a message, and he answers the phone. All this can be difficult to accept if you are used to people punching a clock. But stop and think for a moment. Where is the one place your salesman is not going to sell a building? Your office!

The office is something he will use to help him. It becomes only a back-up operation to his main working area, which is everywhere except the office.

I'll never forget what one of the owners of the firm where I first worked said to me when I started. He told me the one place he'd better not see too much of me was in the office. Neither he nor anyone else in the office had any intention of buying a building from me. So go some place where people might buy.

The truly professional salesman is very independent and wants very little to no supervision. This is even applicable to firms that use a formal sales force, such as the building material manufacturers. Oh sure, they have sales managers, and all the chains of command it takes; but their sales manager knows when he has a professional working for him. He doesn't have to read the paperwork to tell him what the man has been doing; he only needs to glance at the order sheet because that's the only yardstick one can use to measure a professional salesman. Does he get results? That's the bottom line; everything else is just so much window dressing.

A good one-word description of the construction salesman is "maverick." He's a person who is his own man, a self-starter who's good at what he does and knows it. Our construction salesman is the kind whom you tell what you expect to accomplish but not how to do it; that's up to him. These are the conditions he will work best under.

Okay, you say, I get the picture of what I can expect in this individual; but how do I fold him into a construction company operation, and, just as important, where do I find such a person? The rest of this chapter will help with these questions.

INTEGRATING THE SALES DEPARTMENT INTO THE OVERALL OPERATION

Your first step will be the physical part of integrating a sales department into your company. The salesman needs a place where he can take care of paperwork and make phone calls. An office is fine; but if one is not handy and you have a conference room with a phone, there's nothing wrong with using it. Matter of fact, I used a conference room for two years as my office. When the conference room was in use, I took over any empty office. This worked very well, and the firm didn't have to rent more space, which helped on the overhead.

After giving your salesman a place to hang his hat, see that the clerical requirements are covered so that necessary typing and filing will be done. Now with regard to typing, it's important to give the salesman some rank. He needs to have some authority so he can have office help on short notice. And I mean short notice! When the prospect tells your salesman to type up a letter of intent for him to sign, it had better not end up on the bottom of the typing pile. It goes right to the top and is finished promptly.

The salesman has to be able to have a short turnaround time in the office when he needs material to get a job moving. Many times it's simply a case of asking the typist to please type the sales proposal next. In larger offices where there is an office manager, let the salesman go through the manager. The office help technically work for him, and he will have other important work going on; so the salesman should make his request through the proper channels. You have to make clear to the manager, though, that the sales request will be treated as priority when he is asked that it be handled in that manner. Now make sure your salesman understands not to ask for priority treatment unless it's important. Salesmen have a tendency to be impatient when it comes to paperwork, expecting everything to be done for them immediately. There will have to be a little give and take here.

The office help won't understand why this new person has to have a letter typed in the next hour. They are used to having the paperwork move at a set pace. The people doing the work and the people receiving the material all know how the system works. Nothing, they will think, has to be done that quickly; after all, the salesman has been working on that particular prospect for the past four months, and a couple of days certainly won't matter. Well, it just might matter a great deal, especially if it's toward the end of the week. You will have to impress on the office people just how important some of the typing will be.

Even then be prepared to referee every so often; for if your salesman is good, he'll take matters into his own hands when necessary. For example, in 1973 I had been working on one prospect for six months. It was a $200,000 project, and I was determined to sell the job. Well, one Friday morning Mr. X told me to work up a letter of intent and general specifications to go with the drawing I had furnished. I asked if he would be available that afternoon if I could get the papers typed up for his signature. He replied yes, but what was the hurry? I told him I wanted to work on the job over the weekend, which was only partially correct. I also wanted the papers signed before the weekend.

I drove like a maniac back to the office, dashed in, and found no one there except the three women who worked in the office. It was the end of the month, and they all were working on invoices to send out. The office manager was a man who was very sympathetic to the plight of the salesmen (there were two of us) except when he was preparing month-end invoices. Then he guarded the typists' time like a mother hen. There I was with a hot prospect ready to sign, and no one to tell the typists to type for me. And their instructions were implicit: type invoices and nothing else.

Well, I took the attitude that the invoices of sold jobs should take a back seat to one which needed selling. I told two of the typists to stop what they were doing, and I put one to work on the letter of intent (the draft of which I had been carrying around in my briefcase for a month) and the other to typing a specifications list. Just as they were finishing the typing, in walked the office manager. He took one look around and did a marvelous job of holding his temper. I snatched my completed papers, thanked the typists, and bailed out. One hour later I was back with everything signed up.

I also knew I would be hearing from the office manager, and, sure enough, I'd no sooner walked out of the boss's office than he caught me in the hall. I listened politely while he told me that he was going to have to pay overtime to the typists to make up for the time I cost him. He said a few other words which I'll not burden you with. When it was all off his chest, I asked him one question, "What would you have done in my place?" He grinned and was man enough to say, "The same thing you did!"

The point I want to make is that the salesman, if doing his job, will be a disruptive force from time to time.

The next step in bringing this new person into your organization is to explain to all on the nuts-and-bolts side how the salesman fits in. Your superintendents on the jobs will first take the attitude that the salesman will be looking over their shoulder and reporting all back to the boss. The best way to dispel this notion is for the salesman to establish a rapport with the field people. The main reason he should be checking the jobs is to gather information to supply the owner. And let the person running the job know this from both you and the salesman. In the end it'll be up to the salesman to integrate himself into the construction side of the company; and if he's good at his job, he'll do it with ease, since his job is dealing with people.

Now at the same time, if something is discovered that is going to affect the job and create problems with the owner, you'll be the first to know. If the superintendent gets his nose out of joint, then too bad—because your salesman isn't going to stand by and let the owner get the wrong impression of him or the company. And if your salesman doesn't take this attitude, then you don't want him. Aggressive selling means being aggressive in more than one area.

There is a negative aspect to adding a sales department. Hopefully the salesman will be able to make it disappear, but I personally think it never goes away altogether. I'm talking about resentment again, but I feel it needs to be added in this chapter. Every sales force is resented to some degree by the rank and file. By management, no, because they recognize the contribution a sales force makes; but the people sitting behind the desk see a salesman as a free-wheeling good-time Charlie taking a customer to lunch, then writing up an order with commissions amounting to what they make in a couple of months. Then these same people have to take care of the details to make sure the customer receives what he bought.

This resentment can really be present in a construction company with the people working on the site. It can begin with the man in charge and work right on down to the man with a hammer. Therefore, it's imperative to impress on your people in charge of the field just how important a job the salesman has. Stress the concept of teamwork, that somebody has to build what the salesmen sell, and everyone is doing a vitally needed job. As I said, I don't think the resentment will ever go away completely. What you're working toward is making it an insignificant factor and obtaining money-making results from your employees.

It's going to be up to you, the employer, to put the salesman in the picture. Afterwards it'll be his job to make it work. The salesman should report only to you or top management; no one else should have one thing to say about how he does his job. His position is comparable to a staff job where one has the authority to get his work done but has no one answering directly to him.

The main thing you should be interested in is results. The how, you leave to your construction salesman.

SALESMAN'S RESPONSIBILITY TO THE COMPANY

As much of an individual as the salesman will probably be, and as much as he will dislike paperwork, there will be certain duties that you should expect. Proper records are a must. It's vital for you to know what's being worked on and what the near future holds for contracts. The salesman should keep a lead and prospect book with all pertinent information recorded for each listing, status of the prospect at the last meeting, and what is to be done before the next meeting. The time of the next meeting should be recorded also. This book will serve two purposes: give you, the employer, a handy list to check to get a feel for what the salesman is pursuing, and provide the salesman with a formal book of prospects that he can use to keep all the many loose ends under control.

Now a good salesman will always have an up-to-date prospect list. The problem is that it will probably be in a small notebook he carries with him all the time. The office book will require that the information be kept in a more accessible manner.

Along with the prospect book it's very important to spend some time every one to two weeks, whatever works out best, with your salesman and go over the prospects as well as owner relations with jobs under construction. This exchange of information can be very helpful to both of you, and all concerned will stay up-to-date.

One word of caution. Please don't try to tell the salesman how he should go about selling. I've seen it happen. It's hard for employers not to express their ideas on how the salesman should be doing his job. Convey overall expectations, yes; but not the details. That's what you're paying the salesman to know. So fight the urge to do it.

Communications during the day are pretty much up to the salesman. I've found it best that he check with the office at least three times, and never any less than twice. It will be a good idea to inform him that you want him always to take the time to call the office. Now let me say here that if your salesman is a pro, he'll check with the office, since the office will be one of his best sources of leads. This is an excellent yardstick against which to measure your salesman after he starts to work. If he doesn't check the office to pick up messages, start watching his performance. In my opinion, he's not doing his best.

You may be thinking about a pocket beeper for your salesman to use. After all, the other men in the field use them. My objection to the things is that they always seem to go off at an inconvenient time for me. I personally didn't like to have a beeper start to nag me while I was talking with a lead or prospect. Why spend the money if you're not going to use it; so I chose to carry plenty of dimes to call the office. This is strictly my personal preference; other salesmen use them very effectively, I'm sure.

While working, the construction salesman must be on guard against his giving orders to the construction people. Now the pro knows how to handle the construction side of the business; but if you have someone coming on board who doesn't know very much about how a construction outfit operates, then tell him what not to do. This may seem like a strange statement, but bear with me, and my point will be understood. I know personally how difficult it sometimes is to keep quiet. Nevertheless, the salesman should go through the person in charge when he sees something out of line. The main reason is that although the men don't work for him, he is recognized as management; therefore, most likely they will accept orders from the salesman. Thus, the salesman will be undermining the superintendent, which is guaranteed to create friction between the sales and the construction departments. Please be the boss and see that it doesn't happen.

At some time everyone in the company will look on the salesman as an errand boy, even you. It's only natural to say to let so-and-so drop it off. This can get completely out of hand. It is best for all concerned to make it clear that the salesman is not a messenger. Now I'm the first to admit there will be occasional exceptions that are not inconvenient or are important enough to warrant it. What you must be on guard against is letting these exceptions become more frequent. You can tell the office people, and the salesman as well can say no to the office staff; but how about you? You are the one person the salesman works for, and he cannot refuse you; so it's mainly to you that I'm directing this advice. That salesman is worth too much to be used in this manner.

The last suggestion on this subject is not to expect the salesman to keep regular office hours. He must be free to move where the business takes him on the time schedule that gets the job done. As I said before, he won't sell a thing in the office.

FINDING AND TRAINING THE CONSTRUCTION SALESMAN

You're wondering about now just how you locate a construction salesman. You know the local employment agencies won't be able to help you. This is simply too specialized a field. So you're back to square one.

It's not really that difficult if you know where and for what to look. Here I can help you. Right off the bat you can check around with the competition and determine who's using salesmen. If there are some operating in your area, then you can approach them to see if any will consider working for you. This is probably the fastest way to hire a salesman. It can also be the most costly in money; don't expect to bring in an experienced person for peanuts. The real problem is that there are not very many truly professional construction salesmen around; so chances are you'll have to create one.

Now if you list the people working for you, to think about whom you can make a salesman, you're just wasting your time. You might luck out and have someone on your payroll who can do the job, but it's a million-in-one shot. People with sales ability don't go to work for construction firms; nuts-and-bolts people work for contractors. Sales-minded people go where salesmen are used; so that's where you have to look.

The very best place to start your hunting expedition is within the construction industry itself as well as closely allied fields. I really don't recommend going outside this area because then the salesman will know absolutely nothing of the construction industry.

You're looking for a person who's first and foremost a salesman; that's the most important fact. Next you want this salesman to have a little knowledge of construction. He doesn't have to be an ex-superintendent, but it helps if he knows the difference between a concrete slab and a built-up roof. You can teach a good salesman the necessary construction details to become a successful construction salesman. After all, salesmanship is what the salesman is bringing with him; the construction side is your area.

Now this salesman must be used to selling to the person in charge. An order taker who calls on purchasing agents is not for you. Your salesman must understand why he has to deal with the person who signs the checks. Next it's very helpful if his background has been in larger-amount orders—I'm talking about experience in orders for thousands of dollars, not a bunch of nickel-and-dime items.

You want a person who's been in your area long enough to get established. So much of construction selling is who you know, and a newcomer will be at a decided disadvantage, so much so that he may never get off the ground. Please keep in mind the main theme: the person must be a salesman. A pro will pick up the details of the job quickly; but if you have to train a salesman to sell, then you're wasting your time.

Age is a factor that will affect sales. I have a stockbroker friend who was delighted when he started getting gray at the temples. He also tells the story about another broker who had a little gray added over his ears to look older. Whether the story is true or not, the point is well taken. There are several selling positions where maturity is a definite advantage, and construction sales is right at the top of the list. By far, the owners in the market for a new facility will be around forty or older. These owners will be graduates of the business school of hard knocks, and they may hesitate to deal with what to them must seem a kid. Confidence is a large slice of the selling pie, and these owners when talking about thousands of dollars prefer to talk with mature people. In my estimation the construction salesman should look thirty-five or older. The old adage, "Don't send a boy to do a man's job," has never been more true than when applied to construction sales.

The whole thing boils down to this: you're looking for a mature, established professional salesman who has experience selling construction-related items in large-amount orders to the person who signs the check. And believe it or not, there are plenty of candidates running around in the construction industry and closely allied fields.

Such as who, you ask? Well, for openers, there are the factory reps for the building material manufacturers. They fit the above description to a T. Factory reps for the pre-engineered metal building manufacturers also fall into this category. Construction equipment salesmen are another good source. Any segment of the industry that uses salesmen can offer the good raw material for a construction salesman.

Then you have the allied fields. Again there are the factory reps, for construction steel and other metal products; industrial mill supply salesmen will have an acceptable background. There are dozens of areas that offer good salesman prospects, which will come to light when you begin to dig around.

It's my opinion that the factory reps for building materials and metal building manufacturers are the most suitable source of good construction salesmen. They will have all the assets plus the added advantage of being used to running a sales territory on their own. To them, the home office is not across town but in the next state. To do a successful job, they have to be self-starters and able to work with very little supervision. Both of these qualities are absolutely essential to the results-getting construction salesman. The only weakness with these reps is being established in your area. Don't consider anyone unless he does make his home in your working area. If you happen to find a man who's a hometown boy, then so much the better.

When you find one of these reps who does want to make a change, do your homework to find out why. You don't want to end up with some sales manager's problem child. You'll discover after talking with some of these people about changing jobs that the main reasons they are thinking about leaving are desire to get off the road and wanting better opportunities to make more money. This will come home to them while they are in their thirties; so their age is in your favor also.

The other areas I've mentioned are good salesman sources also, only not quite so good as the factory rep group; don't ignore these other sources. Matter of fact, one of the best construction salesmen I ever had the pleasure of knowing and working with came from a supplier of metal products that touched only slightly on the construction industry. He was a prime example of why you have to hire a salesman first, with all else secondary. This guy was the best psychologist I've ever seen when it came to the prospect. He knew exactly what the prospect was going to say before he opened his mouth. He knew how to approach the job, and, better yet, he was a closer—he got the prospect's name on the dotted line. But he always had a little trouble with construction details. He never learned the correct names for all the details in some areas, such as electrical. For example, everything was a switch or a plug on the wall; never exactly what it was, simply a switch or a plug. He never could remember the difference between a column and a beam; and I'll bet right now if you asked him, he'd stop and think about it and then give you the wrong answer. After a while the two owners of the construction company that we both worked for gave up on him. If the prospects didn't mind, why should they? He just kept bringing in the contracts.

He was a fine salesman, always selling himself and his company, and on the lookout for leads twenty-four hours a day, seven days a week. This man was a super construction salesman who came from a metal products firm. He never mastered all the details on the construction side; and, what's more, he would tell you he didn't care. He was a salesman, not an engineer.

My point is that no matter where you find your salesman, he'd better be a salesman first.

OUTLINING TERRITORIES WHERE TWO OR MORE SALESMEN ARE INVOLVED

If you have or plan to have two or more salesmen, there should be definite areas of responsibility outlined. These established territories should be for leads and prospects, not actual construction sites. The territory is where the salesman will work seeking leads and working on prospects. This includes incoming phone leads from the territory, as well. Now if the prospect is going to build and the site is in another man's territory, this situation will present no problem. The job belongs to the man who sold the job. This particular situation will come up because the prospect will be moving or in some cases adding a branch operation.

There will be conflicts from time to time involving leads that start in one area and end up in another salesman's territory. My personal feeling is that the lead in such a case should be passed on to the man who handles the territory. Or you can remain uninvolved, and let the two men in question work things out. The one thing that you can't afford to allow is to have your salesmen fighting and squabbling over leads. It's important that they work together, for if you're doing any business at all, there'll be times when one has to cover for the other. Good relations in the construction sales department is a must.

You may feel the salesmen are mature and objective enough to work out the territories on their own. If so, let them, as long as the results are there. After all, results are what you're really interested in. My last bit of advice on this subject is not to create a very formal sales department with rigid guidelines. What you want is an efficient, fluid operation that does what has to be done to get the contract. Let me give you a good example. In 1974 I was one of two salesmen working for a construction company. A lead came to my attention that was located in the other salesman's territory. The reason I got the lead was that the party in question was a personal friend. I hesitated to pass the lead on to the other salesman because I was the one who knew the party and the background, and he had come to me because we were friends. Therefore, I was the one with the best chance of having a contract signed. Also it was a lead that was not out on the streets, and chances were the other salesman would never have heard of the job. We talked it over. The first thing was to get the project for the firm; and since I had the inside track, he told me to sell and handle the job, which I did.

Now this is what I mean about an efficient, fluid organization, getting the job done. So don't shackle your salesmen with rigid rules. Let them operate in an independent manner, as long as it gets the results.

SALESMAN REMUNERATION

Remuneration in plain language is *money!* Nothing's free and 99% of the time you get exactly what you pay for. A good results-getting construction salesman is going to cost you. A factory rep will be making around $20,000 at least, not including company car, expense accounts, and fringe benefits. Salesmen with local suppliers may be earning in this neighborhood too (probably a little less from my experience).

I'm certainly not saying for you to go right ahead and put a man on a $20,000-a-year salary to get him started. But there has to be a starting place; so a good rule of thumb is 3 to 3½% of the contract amount with you paying the salesman's expenses. If you furnish a car, then the 3 to 3½% should drop by what you feel a car is worth. At $500,000 worth of business, the salesman will make $15,000 to $17,500. I'm talking about $500,000 worth of business in the area where design-build works best, $20,000 to $300,000; so the half million will most likely represent, say, four to eight contracts. These amounts should represent the minimum and be the starting point.

But look what happens at a million in yearly sales. The salesman then is making a nice income, and that is what he should always be working for. Don't ever even consider putting a limit on what your salesman can make. Let him make $100,000 if he can, since when he makes his, he's making yours.

How you handle the money is really up to you and your salesman. You have several options; a small guaranteed salary, say $500 a month, and 2% of the contract amounts; a draw against commissions; or straight commission. If a man works straight commission and furnishes his own expenses, then he should make no less than 4%. Frankly I don't feel this is the way to go. You're thinking right now that straight commission is something to look into because you have no outlay involved. You're right; and you have very little control over the man either. He'll do just as he pleases because you pay him on results only; if he wants to take off a week, there's not much you can say.

In my opinion the best arrangement is a small salary, say $5,000 to $7,000 a year, and 2 to 2½% on sales. You keep control and the salesman has a chance to make some money. At the half-million figure the salesman will make $15,000 to $19,500, based on the above numbers.

Not being familiar with incomes anywhere except where I work, I don't know if I'm helping very much with what your salesman should make. I believe you'll have to derive this figure from what is accepted practice in your particular area. If there is no accepted practice, then the commission percentages I've listed will make a good starting point. It really boils down to what you and the salesman agree to.

Under no condition should you consider a straight salary. First, if the prospective salesman wants a straight salary, mark him off your list. This is a good indicator of what kind of salesman he is. A straight-salaried man is not aggressive enough for construction sales. It's too easy to become lazy when the money is always there. Of course, if he's not working, the lack of results will show it, and you can take proper action. Trouble is you will be out time and money.

Most of the salesman's money should come from commissions, and the salesman, if he's the caliber you want, will prefer it this way.

I imagine this is all new to you; so rely on your own good common business sense about the salesman's income. Just remember to maintain some control over him, to put no limit on earned income, and never to pay a straight salary.

But it falls into your lap, as the employer, to make things happen first, by creating the sales department in your company.